MW00334577

Transportation Depth Six-Minute Problems
for the PE Civil Exam

Seventh Edition

Norman R. Voigt, PE, PLS

PPI®
PPI2PASS.COM
A **KAPLAN** COMPANY

Register Your Book at ppi2pass.com

- Receive the latest exam news.
- Obtain exclusive exam tips and strategies.
- Receive special discounts.

Report Errors for This Book

PPI is grateful to every reader who notifies us of a possible error. Your feedback allows us to improve the quality and accuracy of our products. Report errata at **ppi2pass.com**.

TRANSPORTATION DEPTH SIX-MINUTE PROBLEMS FOR THE PE CIVIL EXAM
Seventh Edition

Current release of this edition: 3

Release History

date	edition number	revision number	update
Aug 2018	7	1	New edition. Code updates. Copyright update.
Jan 2019	7	2	Minor corrections.
Apr 2019	7	3	Minor corrections. Minor cover update.

PPI
1250 Fifth Avenue, Belmont, CA 94002
(650) 593-9119
ppi2pass.com

ISBN: 978-1-59126-621-1

Table of Contents

TOPIC IX: Drainage

TOPIC X: Alternatives Analysis

About the Author

Norman R. Voigt, PE, PLS, is a registered civil engineer and land surveyor in Pennsylvania. Mr. Voigt obtained his bachelor of science and master of science degrees in civil engineering from the University of Pittsburgh. He has worked in both the private and public sectors, performing design, construction, maintenance, inspection services, and interagency project coordination on a variety of highway and transit projects. He has served as the principal coordinator for the civil PE intensive review course at Penn State Beaver, including organizing the course curriculum and teaching the transportation portion for more than 35 years. He has served as adjunct faculty at the University of Pittsburgh and The Pennsylvania State University. He has presented papers on various transportation subjects and on the role of adjunct faculty in undergraduate education, and has authored and edited several books on transportation engineering topics.

Preface and Acknowledgments

The Principles and Practice of Engineering (PE) examination for civil engineering, prepared by the National Council of Examiners for Engineering and Surveying (NCEES), is developed from sample problems submitted by educators and professional engineers representing consulting, government, and industry. PE exams are designed to test examinees' understanding of both conceptual and practical engineering concepts.

I wrote *Transportation Depth Six-Minute Problems for the PE Civil Exam* to provide problem-solving practice consistent with the NCEES exam specifications for the transportation depth section of the PE civil exam. For the seventh edition of this book, I updated existing problems to the latest design standards and references used on the PE civil exam. The seventh edition is aligned with the topics and distribution of problems on the exam.

I have received many helpful comments about the material, and great support from the PPI staff, so that each subsequent edition has been updated to keep up with expected changes in the exam specifications. Thank you to Grace Wong, director of editorial operations; Cathy Schrott, editorial operations manager; Thomas Bliss, associate production manager; Tom Bergstrom, technical illustrator; Tyler Hayes, senior copy editor; Bradley Burch, production editor; Ellen Nordman, publishing systems specialist; and Sam Webster, publishing systems manager.

Thanks to R. Wane Schneiter, PhD, PE, DEE, and Bruce Wolle, MSE, PE, for contributing water resources and environmental problems and geotechnical problems, respectively, which appear in the Drainage and Geotechnical and Pavement chapters.

I especially want to thank the technical reviewers, Dr. Maher Murad; Jamie Rana, PE, PTOE; and Phil Luna, PE; who technically reviewed problems in the first, third, and fifth editions, respectively. And, of course, the support of my wife, Mary Jean, a professional librarian, has made it possible for me to continue to review, update, and organize my material so that the problems represent what is most essential for PE exam preparation.

Problems and solutions have been carefully prepared and reviewed to ensure that they are appropriate and understandable, and that they were solved correctly. If you find errors or discover an alternative, more efficient way to solve a problem, please bring it to PPI's attention so your suggestions can be incorporated into future editions. You can report errors by logging on to PPI's website at **ppi2pass.com**.

Norman R. Voigt, PE, PLS

Introduction

ABOUT THIS BOOK

Transportation Depth Six-Minute Problems for the PE Civil Exam is organized into ten chapters. Each chapter contains problems that correspond to the format and scope that would be expected in the afternoon transportation depth section of the PE civil exam.

Most of the problems are quantitative, requiring calculations to arrive at a correct solution. A few are non-quantitative. As on the exam, some of this book's problems will require a little more time to solve than others. On average, you should expect to spend about six minutes solving each problem.

Each problem includes a hint to provide direction in solving the problem. The solutions are presented in a step-by-step sequence to help you follow the logical development of the correct solution and to provide examples of how you may want to approach your solutions as you take the PE exam.

In addition to the correct solution, you will find an explanation of the faulty solutions leading to the three incorrect answer options. The incorrect options are intended to represent common mistakes specific to different problem types. These may be simple mathematical errors, such as failing to square a term in an equation, or more serious errors, such as using the wrong equation.

Though you won't encounter problems on the exam exactly like those presented in this book, reviewing these problems and solutions will increase your familiarity with the exam's format, content, and solution methods. This preparation will help you considerably during the exam.

Transportation Depth Six-Minute Problems for the PE Civil Exam includes some problems related directly to geotechnical and water resources engineering. You'll find additional problems from these subdisciplines, which are included in the transportation exam, in *Six-Minute Solutions for Civil PE Exam Geotechnical Depth Problems* and *Six-Minute Solutions for Civil PE Exam Water Resources and Environmental Depth Problems*.

ABOUT THE EXAM

The PE civil exam is made up of 80 problems and is divided into two four-hour sessions. Each session presents 40 multiple-choice problems. Only one of the four options given is correct, and the problems are completely independent of each other.

The morning session of the PE civil exam is a broad exam covering eight areas of general civil engineering knowledge: project planning; means and methods; soil mechanics; structural mechanics; hydraulics and hydrology; geometrics; materials; and site development. All examinees take the same morning exam.

Examinees must choose one of the five afternoon exam sections: construction, geotechnical, structural, transportation, or water resources and environmental. The transportation depth section of the exam is intended to assess your knowledge of transportation engineering principles and practice. The topics and approximate distribution of problems for the transportation depth section are as follows.

1. **Traffic Engineering (Capacity Analysis and Transportation Planning) (11 questions):** Uninterrupted flow; street segment interrupted flow; intersection capacity; traffic analysis; trip generation and traffic impact studies; accident analysis; nonmotorized facilities; traffic forecast; highway safety analysis

2. **Horizontal Design (4 questions):** Basic curve elements; sight distance considerations; superelevation; special horizontal curves

3. **Vertical Design (4 questions):** Vertical curve geometry; stopping and passing sight distance; vertical clearance

4. **Intersection Geometry (4 questions):** Intersection sight distance; interchanges; at-grade intersection layout, including roundabouts

5. **Roadside and Cross-Section Design (4 questions):** Forgiving roadside concepts; barrier design; cross-section elements; Americans with Disabilities Act (ADA) design considerations

6. **Signal Design (3 questions):** Signal timing; signal warrants

7. **Traffic Control Design (3 questions):** Signs and pavement markings; temporary traffic control

8. **Geotechnical and Pavement (4 questions):** Sampling and testing; soil stabilization techniques; settlement and compaction; excavation; embankment; mass balance; design traffic analysis and pavement design procedures; pavement evaluation and maintenance measures

9. **Drainage (2 questions):** Hydrology; hydraulics, including culvert and stormwater collection system design, and open-channel flow

10. **Alternatives Analysis (1 question):** Economic analysis

HOW TO USE THIS BOOK

To optimize your study time and obtain the maximum benefit from these problems, consider the following suggestions.

1. Complete an overall review of the problems and identify the subjects that you are least familiar with. Work a few of these problems to assess your general understanding of the subjects and to identify your strengths and weaknesses.

2. Locate and organize relevant resource materials. (See Design Standards and References in this book as a starting point.) As you work problems, some of these resources will emerge as more useful to you than others. These are what you will want to have on hand when taking the PE exam.

3. Work the problems in one subject area at a time, starting with the subject areas that you have the most difficulty with.

4. When possible, work problems without using the hints. Always attempt your own solutions before looking at the solutions provided in the book. Use the solutions to check your work or to provide guidance in solving the more difficult problems. Use the incorrect solutions to help identify pitfalls and to develop strategies to avoid them.

5. Use each chapter's solutions as a guide to understanding general problem-solving approaches. Although problems identical to those presented in *Transportation Depth Six-Minute Problems for the PE Civil Exam* will not be encountered on the PE exam, the approach to solving problems will be similar.

For further information and tips on how to prepare for the transportation depth section of the PE civil exam, consult the *PE Civil Reference Manual* or PPI's website, **ppi2pass.com**.

Design Standards and References

The information that was used to write and update this book was based on the exam specifications at the time of publication. However, as with engineering practice itself, the PE examination is not always based on the most current design standards or cutting-edge technology. Similarly, design standards and regulations adopted by state and local agencies often lag issuance by several years. It is likely that the design standards that are most current, the design standards that you use in practice, and the design standards that are the basis of your exam will all be different.

PPI lists on its website the dates and editions of the design standards and regulations on which NCEES has announced the PE exams are based (**ppi2pass.com**). It is your responsibility to find out which design standards are relevant to your exam. In the meantime, here are the design standards that have been incorporated into this edition.

DESIGN STANDARDS AND REFERENCES USED ON THE EXAM

AASHTO: *AASHTO Guide for Design of Pavement Structures* (GDPS-4-M), 1993, and 1998 supplement, American Association of State Highway and Transportation Officials, Washington, DC

AASHTO: *A Policy on Geometric Design of Highways and Streets*, 6th ed., 2011 (including November 2013 errata), American Association of State Highway and Transportation Officials, Washington, DC

AASHTO: *Guide for the Planning, Design, and Operation of Pedestrian Facilities*, 1st ed., 2004, American Association of State Highway and Transportation Officials, Washington, DC

AASHTO: *Highway Safety Manual*, 1st ed., 2010, vols. 1–3 (including September 2010, February 2012, and March 2016 errata), American Association of State Highway and Transportation Officials, Washington, DC

AASHTO: *Mechanistic-Empirical Pavement Design Guide: A Manual of Practice*, 2nd ed., July 2015, American Association of State Highway and Transportation Officials, Washington, DC

AASHTO: *Roadside Design Guide*, 4th ed., 2011 (including February 2012 and July 2015 errata), 2011, American Association of State Highway and Transportation Officials, Washington, DC

AI: *The Asphalt Handbook* (MS-4), 7th ed., 2007, Asphalt Institute, Lexington, KY

FHWA: *Hydraulic Design of Highway Culverts*, Hydraulic Design Series No. 5, Publication No. FHWA-HIF-12-026, 3rd ed., April 2012, U.S. Department of Transportation—Federal Highway Administration, Washington, DC

HCM: *Highway Capacity Manual*, 6th ed., Transportation Research Board—National Research Council, Washington, DC

MUTCD: *Manual on Uniform Traffic Control Devices*, 2009 (including Revisions 1 and 2, May 2012), U.S. Department of Transportation—Federal Highway Administration, Washington, DC

PCA: *Design and Control of Concrete Mixtures*, 16th ed., 2016, Portland Cement Association, Skokie, IL

PROWAG: *Proposed Accessibility Guidelines for Pedestrian Facilities in the Public Right-of-Way*, July 26, 2011, and supplemental notice of February 13, 2013, United States Access Board, Washington, DC

RECOMMENDED REFERENCES

You may also find the following references helpful in completing problems in *Transportation Depth Six-Minute Problems for the PE Civil Exam*, as well as during the exam.

Manual of Traffic Signal Design. Institute of Transportation Engineers.

Manual of Transportation Engineering Studies, Institute of Transportation Engineers.

Parking Structures: Planning, Design, Construction, Maintenance and Repair. Chrest, Anthony P., Mary S. Smith, and Sam Bhuyan

Principles of Highway Engineering and Traffic Analysis. Mannering, Fred L., Walter P. Kilareski, and Scott S. Washburn

Route Location and Design. Hickerson, Thomas F.

Route Surveying and Design. Meyer, Carl F., and David Gibson

Traffic Engineering Handbook, Institute of Transportation Engineers, Wolshon, Brian and Anurag Pande

Nomenclature

a	acceleration	ft/sec^2
a	regression coefficient	–
A	area	ft^2
A	number of autos per household	–
A	sinking fund amount	–
A	top dimension	ft
A	total grade change in vertical curve	%
ADT	average daily traffic	vpd
AADT	annual average daily traffic	vpd
b	regression coefficient	–
B	base dimension	ft
BF	bulking factor	decimal
BFFS	base free-flow speed	mph
c	cost of roadway excavation	$/yd^3
C	capacity of roadway approach	vph
C	circumference	in
C	single cycle length	sec
C_d	orifice coefficient	–
C_r	calibration factor for roadway segments	–
CMF	crash modification factor	–
CS	critical sum of flow ratios	–
d	deceleration	ft/sec^2
D	degree of curve	deg
D	density of flow	pcpmpl
D	depth of diagonal parking space	ft
D	diameter	ft
D	distance	ft
D_p	phase duration	sec
e	extension of effective green	sec
e	superelevation rate	decimal, %
e	void ratio	–
E	elevation	ft
E	external distance	ft
E	passenger car equivalent	–
E_T	equivalent number of through cars for each heavy vehicle	–
EAL	equivalent axle loading	kips
f	adjustment factor	decimal
f	coefficient of friction	–
f	pipe friction factor	–

F	fuel consumption	gal/veh-hr
F	future worth	$
F	speed adjustment	mph
FFS	free-flow speed	mph
FS	factor of safety	–
g	gravitational acceleration, 32.2	ft/sec^2
g_e	green extension time	sec
g_s	queue service time	sec
G	grade	decimal, ft/sta, %
GF	growth factor	–
h	altitude of triangular shape	ft
h	cost of overhaul	$/yd^3-sta
h	depth of aggregate layer	in
h	elapsed time between bicycle arrivals	min/bicycle
h	elevation head above the orifice	ft
h_f	head loss in the pipe	ft
H	height	ft
H	number of handicapped parking spaces	–
H	height	ft
HMVM	crashes per 100 million vehicle miles	–
HSO	horizontal sight offset	ft
i	annual growth rate	decimal
I	angle of intersection between two tangents	deg
I	deflection angle	deg
I_w	width of travel aisle	ft
$I_{w,han}$	width of handicapped stall aisle	ft
K	probability factor determined by desired level of significance	–
K	ratio of vertical curve length to grade change	ft/%
l_1	start-up lost time	sec
l_2	clearance lost time	sec
leh	length of economical haul	ft
lfh	length of freehaul	ft
loh	economical length of overhaul	ft
L	length	ft, sta
L	lost time per signal cycle	sec

LEF	load equivalent factor	–
LP	low point of vertical curve	–
m	mass	lbm
M	mid-ordinate	ft
n	number	–
N	north coordinate	ft
N	number	–
N	number of maneuvers per hour	hr^{-1}
O	vehicle occupancy	–
p	offset distance from shifted PC to spiral tangent	ft
p	proportion	–
P	applied load	kips
P	number of persons	–
P	pile load	kips
P	present worth	$
P	proportion of traffic in a certain class	decimal
PC	point of curvature	–
PI	point of intersection	–
PHF	peak hour factor	–
PT	point of tangent	–
PVC	point of vertical curve	–
PVI	point of vertical intersection	–
Q	discharge rate	ft^3/sec
Q	flow rate	ft^3/sec
R	crash rate	–
R	curve radius	ft
R	rate of change in grade	%/sta
s	distance	ft
s	saturation flow rate for lane group	vph
s_0	base saturation flow rate	vph
S	bicycle spacing	–
S	mean speed	mph
S	salvage value	$
S	settlement	in
S	sight distance	ft
S_d	depth of parking stall 90° to aisle	ft
S_l	length of parking stall	ft
S_w	width of parking stall	ft
$S_{w,\text{han}}$	width of handicapped parking stall	ft
t	time	min
T	number of person-trips per house hold per day	–
T	tangent length	ft
T_A	number of auto trips per day	–
TF	truck factor	–
UTF	urban travel factor	–
v	velocity	ft/sec, mph

v	volume	ft^3, yd^3
V	flow rate	vph, pers/hr
V	volume	ft^3, yd^3
V_g	unadjusted demand flow rate	vph
V_p	rate of flow during peak period	pcphpl
VM	vehicle-miles traveled	veh-mi
w	weight	lbf
w	width	ft
W	walkway width	ft
x	horizontal distance from PVC	ft
x	tangent distance	ft
X	degree of saturation for signal phase	decimal
y	critical flow ratio	–
y	offset from tangent	ft
Y	yellow change interval	sec
z_1	upstream manometer reading	ft
z_2	downstream manometer reading	ft
Z	clearance distance	ft

Symbols

α	angle	deg
β	angle	deg
Δ	deflection angle	deg
ϵ	specific roughness	ft
ϵ/D	relative roughness	–
η	efficiency	decimal
μ	mean service rate per server	hr^{-1}
ρ	mass density	lbm/ft^3

Subscripts

3ST	three-leg intersection with minor-road stop control
15	peak 15 minutes
a	allowable or area
accel	acceleration
ave	average
A	number of access points per mile
b	base dimension or braking
bb	bus blockage
B	base width or bus maneuvers
c	carpool or crash rate
comb	combined
crit	critical
C	cut
decel	deceleration
E	effective
f	following
F	fill
g	grade or lane group
gl	lane in the lane group with the highest volume
h	highway
han	handicapped
horiz	horizontal direction
HV	heavy vehicle
i	intersection or parking space type
ind-pile	circumferential perimeter
int	intersection length
l	length
LC	lateral clearance
Lpb	pedestrian-blockage for left turns
LT	left turn
LU	lane utilization
LW	lane width
m	median type or parking maneuvers per hour
maj	major road
max	maximum
min	minimum or minor road
ms	downstream lane blockage
M	median
n	normal
o	initial or office use
O	optimal
p	driver population, parking, or peak
prop	proposed
r	perception-reaction time or retail use
ra	collision types related to lane width
rd	divided roadway segment
rr	railroad curve
rs	roadway segment
R	recreational vehicle
Rpb	pedestrian-blockage for right turns
RT	right turn
s	soil, spiral, or stopping
seg	segment
set	settlement
sov	single occupant vehicle
sp	parking space
spf	safety performance function
t	total conflicts
tr	transition rate
T	trucks and buses
u	utilization
vert	vertical
w	water or width
wz	work zone presence at the intersection
w/s	warehouse/storage use

1 Traffic Engineering

PROBLEM 1

AASHTO's *A Policy on Geometric Design of Highways and Streets* (GDHS) specifies criteria for safe design speed on highway curves. Which of the following criteria normally apply to designing a safe curve?

   I. curve radius

  II. passenger comfort factor

 III. sight distance

 IV. shoulder width

  V. speed-limit posting

 VI. side friction factor

VII. superelevation rate

VIII. weather conditions

   (A) I, II, III, and VII only

   (B) I, III, VII, and VIII only

   (C) I, II, III, VI, VII, and VIII only

   (D) I, II, III, IV, V, and VIII only

Hint: Safe speed on a curve balances the forces needed to keep a vehicle on the roadway against the forces tending to push the vehicle off the roadway.

PROBLEM 2

A four-lane divided highway in a suburban area has the following characteristics.

lane width	10 ft
average grade	<2%
left clearance	4 ft
right clearance	2 ft
percentage of heavy vehicles	7%
access spacing	300 ft
design speed	60 mph
posted speed limit	55 mph
directional design hour volume	2540 vph
peak hour factor (PHF)	0.92

What is the level of service (LOS) of the highway?

   (A) C

   (B) D

   (C) E

   (D) F

Hint: Roadway configuration restrictions affect the free-flow speed (FFS). Design speed can be considered to be the base free-flow speed (BFFS) if there is no information to the contrary.

PROBLEM 3

A freeway in rolling terrain has the following characteristics.

commuter traffic volume (one way)	1970 vph
number of lanes (in each direction)	4
percentage of trucks	3%
percentage of buses	3%
percentage of RVs	1%
peak hour factor (PHF)	0.85

What is most nearly the peak hour flow rate?

   (A) 477 pcphpl

   (B) 580 pcphpl

   (C) 623 pcphpl

   (D) 661 pcphpl

Hint: The peak hour flow rate is the per-lane passenger-car equivalent of the hourly count of total vehicle flow. Use the formulas for equivalent passenger-car flow rates from the *Highway Capacity Manual* (HCM).

PROBLEM 4

A 10 mi section of freeway in rolling terrain within a metropolitan area of 325,000 population has the characteristics listed. Local commuters make up 90% of the traffic, and the remainder is a combination of through traffic and recreational traffic to a nearby national park.

free-flow speed (FFS)	60 mph (measured)
number of lanes (in each direction)	3
lane width	11 ft
right shoulder width	6 ft
percentage of trucks (TT)	3%
percentage of buses	2%
percentage of RVs	1%
number of interchanges/ramps	4
peak hour factor (PHF)	0.94 (measured)
driver population factor	1.00
hourly traffic volume (one way)	5700 vph
maximum grade	5% for 0.125 mi

What is the level of service (LOS) for this section?

- (A) C
- (B) D
- (C) E
- (D) F

Hint: Heavy vehicle factors for RVs are not always included with trucks and buses.

PROBLEM 5

The average number of cars passing a point is 2200 pcphpl. The cars travel at an average speed of 42 mph. If the average length of a car is 19 ft, the distance between the cars is most nearly

- (A) 43 ft
- (B) 82 ft
- (C) 100 ft
- (D) 130 ft

Hint: Look for a common element among the defining units, which are related to speed, time, distance, and the number of cars.

PROBLEM 6

Which of the following criteria are used to specify traffic density as determined by the *Highway Capacity Manual* (HCM)?

- I.   passenger-car equivalents
- II.   number of occupants in a vehicle
- III. vehicle count per unit of time
- IV.   vehicle length
- V.   vehicle weight
- VI.   vehicle spacing

- (A) I, IV, and VI only
- (B) III, IV, and VI only
- (C) III, V, and VI only
- (D) I, II, IV, and VI only

Hint: Traffic density is expressed as the number of passenger cars per mile, per lane of roadway.

PROBLEM 7

The stopping distance for a car traveling at 50 mph is 461 ft, including the distance traveled during a 2.5 sec perception-reaction time. If a car traveling at 60 mph is to stop in the same distance with the same friction factor, what is most nearly the required perception-reaction time?

- (A) 0 sec
- (B) 0.1 sec
- (C) 0.7 sec
- (D) 2 sec

Hint: The stopping distance includes braking distance and perception-reaction distance.

PROBLEM 8

A car traveling at 50 mph is followed by a car traveling at 60 mph. The lead car suddenly brakes to a stop within a 278 ft distance. Both cars have the same stopping friction factor, and the driver of the following car has a perception-reaction time of 2.0 sec. The road is level.

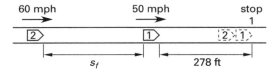

In order to avoid a collision, the minimum distance between the two cars must be most nearly

(A) 86 ft

(B) 280 ft

(C) 300 ft

(D) 320 ft

Hint: The driver of the following car is aware that the lead car is stopping when the driver sees the brake lights turn on.

PROBLEM 9

Passenger stations are spaced 1 mi apart from each other on a rapid transit line. Trains accelerate out of a station at 5.5 ft/sec^2 and decelerate into a station at 4.4 ft/sec^2. Trains have a top speed of 80 mph. What is most nearly the average speed of a train between stations?

(A) 12 mph

(B) 40 mph

(C) 52 mph

(D) 54 mph

Hint: The trip stages between stations consist of accelerating, traveling at constant speed, and decelerating.

PROBLEM 10

A rapid transit line has trains scheduled to arrive at a station every 5 min. The trains each have 4 cars and can carry 220 passengers per car. The trains have a 1 min dwell time at each station. What is most nearly the capacity of the transit line?

(A) 880 passengers/hr

(B) 2640 passengers/hr

(C) 10,600 passengers/hr

(D) 13,200 passengers/hr

Hint: The arrival rate and departure rate are the same, regardless of the length of dwell time at the station.

PROBLEM 11

A high-speed train is to reach 150 mph between stations. Acceleration is limited to 0.18 g, and deceleration is limited to 0.12 g. What is most nearly the minimum spacing between stations?

(A) 0.80 mi

(B) 0.90 mi

(C) 2.0 mi

(D) 6.5 mi

Hint: The train reaches 150 mph for an instant between stations.

PROBLEM 12

In the layout shown, trucks make up less than 1% of vehicles using approach E, and there are 20 parking maneuvers per hour. There are no buses or pedestrians. The location is near a central business district (CBD). The demand flow rate is 1900 pcphpl.

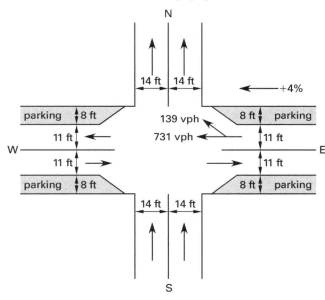

What is most nearly the saturation flow rate of approach E?

(A) 870 vph

(B) 1080 vph

(C) 1180 vph

(D) 1200 vph

Hint: The saturation flow rate modifies an ideal flow rate by approach condition adjustment factors.

Traffic Engineering

PROBLEM 13

Bus service is being planned for an arena event. There will be 12,000 people in attendance, 35% of whom will depart by bus. The buses have an average capacity of 62 persons/bus. The entire arena will empty in one hour at the end of the event. The peak hour factor is projected to be 0.85. Approximately how many buses will be required during the highest peak 15 min after the event?

- (A) 15
- (B) 20
- (C) 70
- (D) 80

Hint: Here, peak hour factor is defined similarly to the highway traffic definition of peak hour factor.

PROBLEM 14

The numbers for airport runways are assigned

- (A) according to the landing heading, or azimuth, rounded to the nearest 10° and dropping the last zero
- (B) according to the take-off heading, rounded to the nearest 10° and dropping the last zero
- (C) based on the airport design firm's internal design procedure
- (D) based on the heading bearing, rounded to the nearest 10°

Hint: The most critical condition is to find a place and direction to land, even under power failure and poor visibility.

PROBLEM 15

A transit station serves a stadium with a capacity of 50,000 persons. When an event is finished, 95% of the people in attendance are expected to leave in one hour, and the transit system is expected to carry 35% of those in attendance. The peak hour factor, PHF, defined similarly to the highway traffic definition of peak hour factor, is 0.75, and the level of service (LOS) is to be D. Most nearly, what is the minimum required effective width of the concourse walkway leading to the station?

- (A) 25 ft
- (B) 34 ft
- (C) 76 ft
- (D) 96 ft

Hint: Use platoon-adjusted LOS criteria.

PROBLEM 16

A parking lot is being constructed for 700 cars near an event site. Event parking will be subject to a fixed fee paid upon entering through a cashier lane. The cashiers' projected service rate, μ, is 270 vph for each entrance lane. During the hour prior to the event start, 72% of the vehicles are expected to arrive. The peak hour factor, PHF, defined similarly to the highway traffic definition of PHF, varies according to the number of lanes, n.

no. of lanes, n	PHF
1	0.70
2	0.80
3	0.87
4	0.94
5 or more	0.97

If V is the arrival rate in vph, and $n = V/\text{PHF}\mu$, most nearly how many entrance lanes should be provided?

- (A) two
- (B) three
- (C) four
- (D) eight

Hint: A fraction of a lane requires the addition of another full lane.

PROBLEM 17

An exclusive bikeway is operating with an average two-way flow of 20 bicycles in the peak 15 min period. The directional distribution is 60%/40%. Most nearly, what is the heaviest one-way density of flow at an average speed of 11.2 mph?

- (A) 0.3 bicycle/mi
- (B) 4.3 bicycles/mi
- (C) 7.2 bicycles/mi
- (D) 27 bicycles/mi

Hint: Density is related to the number of vehicles per unit length of the pathway.

PROBLEM 18

A commuter train stops at stations spaced 1 mi apart. The top speed of the train is 80 mph. The train accelerates at 5.5 ft/sec^2 and decelerates at 4.5 ft/sec^2. It dwells at the platform for 20 sec to pick up passengers. What is most nearly the average running speed of the train?

(A) 33 mph

(B) 41 mph

(C) 52 mph

(D) 64 mph

Hint: Running speed applies to the time during which the train is in motion between stations.

PROBLEM 19

The light-rail vehicles that travel along a 10 mi long line currently make an average of four stops per mile. Due to complaints of slow service, the number of stops is being reduced to three stops per mile. The vehicles accelerate at 3.5 mph/sec and decelerate at 2.5 mph/sec. The normal constant running speed is 35 mph. Each stop increases travel time by 30 sec, which includes deceleration, dwell, and acceleration times. Considering the reduced number of stops, what is most nearly the average increase in vehicle speed?

(A) 0.50 mph

(B) 2.5 mph

(C) 16 mph

(D) 19 mph

Hint: The average speed with no stops is the same as the constant running speed of 35 mph.

PROBLEM 20

Determining service flow rate of a roadway requires which of the following criteria, according to the *Highway Capacity Manual* (HCM)?

    I. driver population factor

    II. heavy vehicle factor

    III. lane width factor

    IV. number of lanes

    V. parking adjustment factor

    VI. peak-hour factor

    VII. average control delay per vehicle

    VIII. volume-over-capacity ratio

(A) I, IV, and VIII only

(B) I, II, IV, and VI only

(C) I, II, III, IV, and V only

(D) II, III, V, VI, and VII only

Hint: Service flow rate is expressed as the number of passenger cars per hour, per lane of roadway.

PROBLEM 21

On freeways, according to the *Highway Capacity Manual* (HCM), which of the following statements is TRUE?

(A) On steep downgrades, trucks have lower passenger-car equivalents than they do on slight downgrades.

(B) On upgrades, higher percentages of trucks tend to have lower passenger-car equivalents.

(C) Trucks are usually a higher percentage of peak hour traffic than of off-peak traffic.

(D) The combined percentage of RVs and lighter weight single-unit trucks is most always greater than the combined percentage of heavy single-unit trucks and tractor-trailers.

Hint: A grouping of trucks on freeways tends to form a platoon, and the trucks tend to operate more uniformly than do passenger cars.

PROBLEM 22

In the *Highway Capacity Manual* (HCM), the peak hour factor is determined by the peak

(A) 15 minutes of flow divided by the peak one-hour flow

(B) hour flow divided by the maximum rate of flow that occurs in the peak interval (usually five minutes)

(C) interval of flow (usually five minutes) divided by the peak one-hour flow

(D) hour flow divided by the peak 15-minute flow rate, expressed as an hourly rate

Hint: The traffic flow rate varies throughout a 1 hr period and usually requires more than 5 min to adjust to a new rate.

PROBLEM 23

When describing highway traffic flow, which of the following statements is NOT true?

(A) A highway with buses in the traffic stream has a higher capacity of persons per hour than does a highway with only automobiles in the traffic stream.

(B) A highway with trucks in the traffic stream has a lower capacity of persons per hour than does a highway with only automobiles in the traffic stream.

(C) For a highway in mountainous terrain with long grades, the same passenger-car equivalent is used for trucks on upgrades as is used for trucks on downgrades.

(D) For a highway with a large number of trucks in the traffic stream (as compared to RVs), the RVs can be combined with the trucks when determining passenger-car equivalents of heavy vehicles.

Hint: Buses and passenger vans increase the person capacity of a highway, while trucks and RVs decrease the person capacity.

PROBLEM 24

Transit planners are attempting to determine expected transit usage from a community that produces 12,000 trips/day. The population density is 10,000 persons/mi^2, and there is an average of 0.80 autos/household. The following model for the urban travel factor (UTF) has been established for the community.

$$\text{UTF} = \left(\frac{1}{1000}\right)\left(\frac{\text{households}}{\text{auto}}\right)\left(\frac{\text{persons}}{\text{mi}^2}\right)$$

$$\% \text{ transit usage} = \frac{\text{UTF}}{0.6}$$

Approximately how many residents can be expected to use transit?

(A) 1500 persons

(B) 1600 persons

(C) 2100 persons

(D) 2500 persons

Hint: Households per auto is the inverse of autos per household.

PROBLEM 25

7000 persons commute daily from a bedroom community to an employment center, with an average commute distance of 8 mi in one direction. The average speed of the commute is 20 mph. 25% of the commuters carpool, and the average carpool has 2.10 persons/veh. Fuel consumption is measured by

$$F = 0.0362 \, \frac{\text{gal}}{\text{veh-mi}} + \frac{0.746 \, \dfrac{\text{gal}}{\text{veh-hr}}}{\text{v}}$$

With a fuel heating value of 125,000 Btu/gal, what is the approximate daily round trip Btu consumption for this commute pattern?

(A) 7.2×10^3 Btu

(B) 4.5×10^8 Btu

(C) 8.9×10^8 Btu

(D) 10×10^8 Btu

Hint: A daily commute includes two trips.

PROBLEM 26

A transit terminal is being developed in an urban neighborhood. One of the goals of site impact analysis is to improve public safety. Which of the following would have the SMALLEST impact on this goal?

(A) improving personal security of urban travelers

(B) improving reliability of transit service

(C) providing adequate lighting levels throughout walkway and platform areas

(D) reducing noise and vibration impacts

Hint: Public safety involves minimizing anxiety about personal safety.

PROBLEM 27

A study zone contains 600 households, each averaging 3.5 persons and 2.2 autos. The modal split is 0.94/0.05 auto to transit, with 0.01 assigned to other modes. The following model has been determined to show the relationship for the number of trips per household.

$$T = 0.78 + 1.3P + 2.3A$$

T is the number of daily trips per household, P is the number of persons per household, and A is the number of autos per household. Approximately how many auto trips per day are generated by the study zone?

(A) 10 trips/day

(B) 5900 trips/day

(C) 6200 trips/day

(D) 98×10^3 trips/day

Hint: The number of auto trips is a portion of the total number of household trips.

PROBLEM 28

On a highway facility, how does the observed hourly vehicle volume differ from the design peak-period flow rate?

(A) The highest 15 min vehicle volume is divided by the highest 1 hr flow rate to obtain the peak-period flow rate.

(B) The observed hourly vehicle volume includes a mix of heavy vehicles, while the design peak-period flow rate has been adjusted for passenger-car equivalents of heavy vehicles, the peak hour factor, the driver population, and the number of lanes.

(C) The observed hourly vehicle volume is divided by the number of observation hours to obtain the peak-period flow rate.

(D) The observed vehicle volume is divided by the number of observation hours and the number of lanes over which the observation took place.

Hint: Various vehicle sizes must be converted to a common unit of vehicle measure.

PROBLEM 29

Each lane of a four-lane freeway has a directional capacity of 2100 vph. The normal directional flow is 3100 vph. An incident blocks one lane for 20 min and then is cleared to allow the full traffic capacity flow.

Approximately how long does it take to dissipate the queue after the blockage has been cleared?

(A) 15 min

(B) 30 min

(C) 50 min

(D) 57 min

Hint: Departure from the blockage is at two rates, while arrival continues at the same rate as before the blockage.

PROBLEM 30

A car brakes suddenly and skids to a stop from 60 mph. The car initially skids 150 ft on pavement with a friction factor of 0.30. The skid continues onto wet grass on hard soil with a friction factor of 0.10. Both parts of the skid are on a 3% upgrade. Approximately how long is the skid on the grassy surface?

(A) 540 ft

(B) 750 ft

(C) 1300 ft

(D) 1600 ft

Hint: An upgrade decreases the length of skid from a given speed.

PROBLEM 31

A car traveling at 70 mph on a 5% downgrade skids 350 ft before striking a retaining wall head-on. The coefficient of friction between the tires and the road is 0.30. What was the approximate speed of the car at impact?

(A) 35 mph

(B) 42 mph

(C) 48 mph

(D) 51 mph

Hint: A downgrade increases the required stopping distance for a given speed.

PROBLEM 32

A car traveling on a 3% upgrade at 60 mph in a construction area skids into a stack of concrete barriers. Skid marks leading to the crash measure 150 ft long. The pavement has a friction factor of 0.30.

Approximately how much farther would the car have skidded if it had not struck the barriers?

(A) 19 ft

(B) 210 ft

(C) 250 ft

(D) 290 ft

Hint: The friction factor is assumed to be an average over the entire skid distance.

PROBLEM 33

A car traveling on a 6% downgrade skids 60 m before colliding with another car. Police on the scene estimated the impact speed at 40 kph and determined that the pavement friction factor was 0.48. What was most nearly the car's speed at the beginning of the skid?

(A) 25 kph

(B) 70 kph

(C) 90 kph

(D) 95 kph

Hint: What additional distance would be required for the car to skid to a stop without a collision?

PROBLEM 34

A car skids 185 ft down a 3% grade to a stop. Friction factors are given as follows.

v_0 (mph)			
30	40	50	60
f 0.59	0.51	0.45	0.35

What was most nearly the speed of the car at the beginning of the skid?

(A) 39 mph

(B) 49 mph

(C) 50 mph

(D) 71 mph

Hint: Friction factors are given as averages based on the initial speed of the skid and are assumed to vary linearly.

PROBLEM 35

Crash data has been tabulated for a 20 mi section of an arterial highway.

	2009	2010	2011	2012
crash type				
fatal	0	3	2	5
personal injury	35	35	50	50
property damage only	130	180	190	230
average daily traffic (ADT) (veh)	15,500	16,000	16,500	17,000

What is most nearly the crash rate for injury and fatal crashes per 100 million vehicle miles (HMVM)?

(A) 0.038 crash/HMVM

(B) 36 crashes/HMVM

(C) 38 crashes/HMVM

(D) 190 crashes/HMVM

Hint: The ADT is defined as the total traffic count per 365 days.

PROBLEM 36

Two lanes of a freeway have a capacity of 4200 vph. The normal flow on these two lanes is 3100 vph. An incident blocks both lanes for 10 min. After 10 min, both lanes are opened to full traffic flow. Approximately how long does it take to dissipate the queue that resulted from the blockage?

(A) 7.0 min

(B) 28 min

(C) 38 min

(D) 67 min

Hint: Departure from the blockage site is at full capacity, while arrival continues at the same rate as before the blockage.

PROBLEM 37

The Federal Highway Administration (FHWA) and National Highway Traffic Safety Administration (NHTSA) have made efforts to encourage states to increase emphasis on and find support for safety countermeasures. Which of the following actions would NOT be part of such a program?

(A) establishing pedestrian safety as a priority

(B) improving ergonomic design of vehicle interiors

(C) increasing moving violation enforcement through work zones

(D) providing improved signing, marking, and delineation

Hint: FHWA and NHTSA deal primarily with on-highway activities.

PROBLEM 38

A signalized intersection has had a large number of crashes due to slippery surface conditions. Which of the following countermeasures would have the greatest likelihood of reducing crashes?

(A) improving roadway lighting

(B) installing advance overhead signals

(C) prohibiting turns

(D) reducing the speed limit on approaches

Hint: Poor surface conditions result in reduced traction for turning and stopping.

PROBLEM 39

A highway intersection with a crossroad has been the site of many left-turn and head-on collisions recently. What is the LEAST likely cause of the crashes?

(A) inadequate gaps in traffic

(B) inadequate roadway lighting

(C) the absence of a special left-turning phase

(D) the large volume of left turns

Hint: Left turns involve more critical space allocation than does any other intersection move.

PROBLEM 40

The following list of roadway segment deficiencies has been prepared for a Transportation Systems Management (TSM) study.

   I. lane width under 11 ft

  II. signs and pavement markings not in conformance with Institute of Transportation Engineers/American Association of State Highway and Transportation Officials (ITE/AASHTO) standards

 III. peak average travel speed less than 20 mph

 IV. roadway drainage does not meet Federal Highway Administration (FHWA) standards

  V. sight distance not in conformance with the *Manual of Uniform Traffic Control Devices* (MUTCD)

 VI. roadway level of service lower than C

 VII. roadway lighting levels below Illuminating Engineering Society (IES) standards

VIII. high roadway crash rates

 IX. too many delivery truck companies servicing business establishments

Which of the deficiencies are LEAST likely to qualify for TSM improvement programs?

(A) III, VII, and IX only

(B) IV, VII, and IX only

(C) I, V, VI, and VIII only

(D) II, V, VII, and IX only

Hint: TSM programs create efficient use of road space through traffic management programs.

PROBLEM 41

A major highway segment that passes through a 2.5 mi business district is being evaluated for crash rates over a 4 yr period. The average daily traffic (ADT) is 28,000 vpd. During the study period, there have been 9 crashes/yr involving death or injury. The statewide statistic for similar types of roadways is 150 crashes per 100 million vehicle miles (HMVM) involving death or injury. The business district evaluation uses a rate quality control method in which the critical crash rate, R, for a segment is determined by the following equation, where all variables except K have the same units.

$$R_{\text{crit}} = R_{\text{ave}} + K \sqrt{\frac{R_{\text{ave}}}{\text{traffic base}}}$$

A confidence level of 95% is being assigned to the study, using a K value of 1.645.

The relationship appoximates one standard error, and the units do not necessarily balance. However, the crash rate and the traffic base use the same magnitude of vehicle miles. What is most nearly the ratio of the crash rate of this segment to the statewide critical rate?

- (A)   0.14
- (B)   0.19
- (C)   0.88
- (D)   5.4

Hint: The confidence level is selected for the probability that the crash rate is more than a random occurrence.

PROBLEM 42

Using *Highway Safety Manual* (HSM) procedures, what is most nearly the predicted average crash frequency per year for a segment of multilane rural highway with the characteristics shown?

segment length	2 mi
annual average daily traffic (AADT)	18,000 veh/day
lane width	11 ft
number of lanes	4
right shoulder	6 ft paved
median	8 ft traversable
median barrier	none
lighting	none
local calibration factor, C_r	0.94

- (A)   3.61 crashes/yr
- (B)   3.84 crashes/yr
- (C)   7.23 crashes/yr
- (D)   7.69 crashes/yr

Hint: Crash prediction uses safety performance functions (SPF), such as regression equations, to estimate the average frequency of crashes, which are adjusted according to the site location.

PROBLEM 43

A minor rural two-lane road approaches a major through road to form a T-intersection. The intersection has the following geometric characteristics.

intersection type	3 leg
traffic control	stop control on minor leg
right turn lanes on major road	none
left turn lanes on major road	none
turning lanes on minor road	none
skew angle	0°
approach grades	<1.5%
major road speed limit	50 mph
intersection lighting	none
major road annual average daily traffic (AADT)	7000 veh/day
minor road AADT	1000 veh/day

The crash rate data over the past eight years is within 5% of the statewide mean for intersections of similar configuration.

Using *Highway Safety Manual* (HSM) procedures, what is most nearly the predicted maximum crash frequency for this intersection for the upcoming year?

- (A)   0.467 crashes/yr
- (B)   0.990 crashes/yr
- (C)   1.68 crashes/yr
- (D)   1.77 crashes/yr

Hint: Use the appropriate calibration factor for the location.

SOLUTION 1

Curve radius and speed together determine the radial force needed to hold a vehicle on the curve.

Passenger comfort is important so that the driver can maintain control of the vehicle and so that passengers are not subject to unnecessary disorientation or the feeling of danger.

Sight distance must be adequate so that the driver can avoid hazards.

Side friction must be considered to ensure that the driver can steer the vehicle around the curve.

The superelevation rate makes use of the force of gravity to help keep the vehicle from sliding off the outside of the curve.

Weather conditions such as rain or snow reduce the tire friction available for holding the vehicle on the curve.

The answer is (C).

Why Other Options Are Wrong

(A) This answer omits weather conditions, such as rain or snow, which reduce the available tire friction needed to hold the vehicle on a curve. In regions that are often wet or icy, the design speed is set lower than in regions that have normally dry conditions.

(B) This answer omits criterion II, passenger comfort, which is necessary so that the driver can maintain control of the vehicle and so that passengers are not subject to disorientation or a feeling of danger.

(D) The posted speed limit, criterion V, is subject to local regulatory conditions and does not determine design speed of an existing highway. Posted speed can, however, be used to set the minimum design speed of a new or reconstructed highway.

SOLUTION 2

A criterion of LOS is the maximum service flow rate per hour per lane of highway. The service flow rate is the passenger-car equivalent flow at a free-flow speed. The flow rate must be determined in passenger-car equivalents of the total traffic vehicle mix using *Highway Capacity Manual* (HCM) Eq. 12-9.

$$v_p = \frac{V}{PHF \times N \times f_{HV}}$$

Determine the heavy vehicle factor using HCM Eq. 12-10. HCM Exh. 12-25 shows passenger car equivalents for vehicles in general terrain segments. Heavy vehicles

include all single-unit trucks (SUTs), trailer trucks (TTs), and recreational vehicles (RVs). P_T is the decimal proportion of the total of these three groups.

$$f_{HV} = \frac{1}{1 + P_T(E_T - 1)}$$
$$= \frac{1}{1 + (0.07)(2.0 - 1)}$$
$$= 0.935$$

Determine the service flow rate. f_p is assumed to be 1.0 and is not included in the volume calculation.

$$v_p = \frac{2540 \ \dfrac{\text{veh}}{\text{hr}}}{(0.92)(2 \text{ lanes})(0.935)}$$
$$= 1476 \text{ pcphpl}$$

Determine the adjusted free-flow speed using HCM Eq. 12-3, based on the characteristics shown.

$$FFS = BFFS - f_{LW} - f_{LC} - f_M - f_A$$

The design speed is 60 mph, and the posted speed limit is 55 mph. Therefore, the base free-flow speed (BFFS) is 60 mph (HCM Ch. 12).

The lane width is 10 ft. Therefore, from HCM Exh. 12-20, $f_{LW} = 6.6$ mph.

The total lateral clearance (TLC) is 6 ft. Therefore, from HCM Exh. 12-22, $f_{LC} = 1.3$ mph.

The median is divided. Therefore, from HCM Exh. 12-23, $f_M = 0.00$ mph.

There are access points every 300 ft, which equates to 17.6 access points per mile. Considering a speed reduction of 0.25 mph, or interpolating from HCM Exh. 12-24, $f_A = 4.4$ mph.

$$FFS = 60 \ \frac{\text{mi}}{\text{hr}} - 6.6 \ \frac{\text{mi}}{\text{hr}} - 1.3 \ \frac{\text{mi}}{\text{hr}}$$
$$- 0 \ \frac{\text{mi}}{\text{hr}} - 4.4 \ \frac{\text{mi}}{\text{hr}}$$
$$= 47.7 \text{ mph}$$

The density of flow is found using HCM Eq. 12-11.

$$D = \frac{v_p}{S} = \frac{1476 \ \dfrac{\text{pc}}{\text{hr-ln}}}{47.7 \ \dfrac{\text{mi}}{\text{hr}}} = 30.9 \text{ pcpmpl}$$

The LOS is D.

The answer is (B).

Why Other Options Are Wrong

(A) This answer is incorrect because a service flow rate of 1429 pcphpl would equate to a density of 23.8 pcpmpl, resulting in LOS C at a free-flow speed (FFS) of 60 mph. This answer could only be selected if FFS were not reduced by the flow friction factors.

(C) This answer is incorrect because it assumes a PHF of 0.80, typical of traffic flows near peak demand locations. The resulting density would appear to be LOS E.

(D) This answer is incorrect because not dividing the adjusted flow volume by two lanes results in a density that appears to be LOS F, or jam density.

SOLUTION 3

The total volume, which consists of a mix of vehicle types, must be converted to equivalent passenger-car volume by assigning passenger-car equivalents to the trucks, buses, and RVs. The heavy vehicle factor, f_{HV}, is determined by *Highway Capacity Manual* (HCM) Eq. 12-10. (See Exh. 12-25 for passenger-car equivalents.)

Heavy vehicles include all single-unit trucks (SUTs), trailer trucks (TTs), and recreational vehicles (RVs). P_T is the decimal proportion of the total of these three groups.

$$f_{HV} = \frac{1}{1 + P_T(E_T - 1)}$$
$$= \frac{1}{1 + (0.07)(3.0 - 1)}$$
$$= 0.877$$

The equivalent passenger-car flow rate is determined by HCM Eq. 12-9. f_p is assumed to be 1.0 and is not included in the volume calculation.

$$v_p = \frac{V}{PHF \times N \times f_{HV}} = \frac{1970 \, \frac{veh}{hr}}{(0.85)(4 \, lanes)(0.877)}$$
$$= 661 \, pcphpl$$

The answer is (D).

Why Other Options Are Wrong

(A) This incorrect answer results from misplacing the PHF in the numerator.

(B) This erroneous answer occurs if the correction for passenger-car equivalents was not included in the volume adjustment.

(C) This erroneous answer occurs if the passenger-car equivalents were selected for level terrain instead of for rolling terrain.

SOLUTION 4

The free-flow speed, FFS, is given as

$$FFS = 60 \, mi/hr$$

RVs (1%) and buses (2%) are counted together as single-unit trucks (SUTs), for a total of 3%. With trailer trucks (TTs) at 3% and SUTs at 3%, the mix is considered 50% SUT and 50% TT. Therefore, HCM Exh. 12-27 applies. With 5% grade for 0.125 mi,

$$E_T = 2.25$$

Using HCM Eq. 12-10, determine the heavy vehicle factor, f_{HV}.

$$f_{HV} = \frac{1}{1 + P_T(E_T - 1)}$$
$$= \frac{1}{1 + (0.06)(2.25 - 1)}$$
$$= 0.925$$

Since the driver population is primarily local commuters familiar with the freeway segment, the driver population factor, f_p, is assumed to be 1.0 and is not included in the volume calculation. Determine the peak traffic volume, v_p, using HCM Eq. 12-9.

$$v_p = \frac{V}{PHF \times N \times f_{HV}} = \frac{5700 \, \frac{veh}{hr}}{(0.94)(3)(0.925)}$$
$$= 2185 \, pcphpl$$

Using HCM Eq. 12-11, determine the density, D, in passenger cars per mile per lane.

$$D = \frac{v_p}{S} = \frac{2185 \, pcphpl}{60 \, \frac{mi}{hr}} = 36.4 \, pcpmpl$$

From HCM Exh. 12-15, based on the density range of 35–45 pcpmpl, the LOS is E.

The answer is (C).

Why Other Options Are Wrong

(A) This incorrect answer could result from improper use of PHF and passenger car equivalents. According to HCM Exh. 12-15, in order to operate at LOS C, there would have to be between 18 pcpmpl and 26 pcpmpl.

(B) In order to operate at LOS D, the maximum density would have to have been found to be between 26 pcpmpl and 35 pcpmpl. This density can be incorrectly found by not including the heavy vehicle and driver population factors.

(D) The maximum density for LOS E is 45 pcpmpl. Should the setting be incorrectly interpreted as rural because of the nearby recreational park and the PHF selected as 0.80, the density would be determined to be LOS F.

SOLUTION 5

To obtain the average distance between cars, calculate the number of cars that will be in one mile-lane of roadway.

Calculate the density.

$$D = \frac{V_p}{S} = \frac{2200 \ \frac{\text{pc}}{\text{hr-ln}}}{42 \ \frac{\text{mi}}{\text{hr}}}$$

$$= 52.4 \text{ pcpmpl}$$

Find the spacing between cars in one lane.

$$\text{spacing} = \frac{1 \text{ mi-lane}}{D} = \frac{5280 \ \frac{\text{ft}}{\text{mi-lane}}}{52.4 \ \frac{\text{pc}}{\text{mi-lane}}} = 100.8 \text{ ft/pc}$$

$$\begin{aligned} \text{distance between cars} &= \text{spacing} - \text{average car length} \\ &= 100.8 \text{ ft} - 19 \text{ ft} \\ &= 81.8 \text{ ft} \quad (82 \text{ ft}) \end{aligned}$$

The answer is (B).

Alternate Solution

$$\text{spacing} = \frac{1 \text{ mi-lane} - DL}{D}$$

$$= \frac{5280 \ \frac{\text{ft}}{\text{mi-lane}} - \left(52.4 \ \frac{\text{pc}}{\text{mi-lane}}\right)\left(19 \ \frac{\text{ft}}{\text{pc}}\right)}{52.4 \ \frac{\text{pc}}{\text{mi-lane}}}$$

$$= 81.8 \text{ ft} \quad (82 \text{ ft})$$

Why Other Options Are Wrong

(A) This incorrect answer assumes it is necessary to convert speed to feet per second, then directly subtracts the car length. This gives the length of travel not occupied by a car in one second instead of the actual distance between cars. It is necessary to carry the solution further to determine the following distance using a density relationship.

(C) This answer is incorrect because 101 ft is the spacing from the front of one car to the front of the next car. The length of the car must be subtracted from the spacing to find the distance between cars.

(D) This answer is incorrect because an extra conversion from miles per hour to feet per second was inserted in the first equation.

SOLUTION 6

A common unit of vehicle measure is necessary to make traffic evaluations; therefore, passenger-car equivalents, option I, is chosen as the standard of reference.

The length of the standard reference vehicle, option IV, determines how much roadway surface is occupied by vehicles.

Vehicle spacing, option VI, determines how many vehicles occupy a unit length of roadway, such as one lane-mile, at an instant of time.

The answer is (A).

Why Other Options Are Wrong

(B) Vehicle count data, option III, must be converted to passenger-car equivalents.

(C) Vehicle count data, option III, must be converted to passenger-car equivalents, and the weight of vehicles, option V, is not considered in problems addressing traffic density.

(D) Density, as a definition of vehicle occupancy of a roadway lane, is independent of the number of vehicle occupants, so option II does not apply.

SOLUTION 7

To determine the braking distance, s_b, subtract the perception-reaction distance, s_r, from the total stopping distance, s_s.

$$s_b = s_s - s_r$$
$$s_r = \text{v}t$$
$$= \frac{\left(50 \ \frac{\text{mi}}{\text{hr}}\right)\left(5280 \ \frac{\text{ft}}{\text{mi}}\right)(2.5 \text{ sec})}{3600 \ \frac{\text{sec}}{\text{hr}}}$$
$$= 183.3 \text{ ft}$$
$$s_b = 461 \text{ ft} - 183.3 \text{ ft}$$
$$= 277.7 \text{ ft}$$

The friction factor, f, is found from the formula for deceleration distance.

$$s_b = \frac{\text{v}_1^2 - \text{v}_2^2}{2g(f+G)}$$

Rearrange to find f, setting v_2 at 0 mph, G at 0 ft/ft, and g at 32.2 ft/sec^2.

$$f = \frac{v_1^2}{2gs_b}$$

$$= \frac{\left(\left(50 \ \frac{mi}{hr}\right)\left(5280 \ \frac{ft}{mi}\right)\right)^2}{(2)\left(32.2 \ \frac{ft}{sec^2}\right)(277.7 \ ft)\left(3600 \ \frac{sec}{hr}\right)^2}$$

$$= 0.30$$

Determine the braking distance from 60 mph using the same friction factor.

$$s_{b,60} = \frac{v_1^2 - v_2^2}{2g(f+G)}$$

$$= \left(\frac{\left(60 \ \frac{mi}{hr}\right)^2 - \left(0 \ \frac{mi}{hr}\right)}{(2)\left(32.2 \ \frac{ft}{sec^2}\right)(0.30)}\right)\left(\frac{5280 \ \frac{ft}{mi}}{3600 \ \frac{sec}{hr}}\right)^2$$

$$= 400.3 \ ft$$

Determine the perception-reaction distance for 60 mph.

$$s_{r,60} = s_{s,60} - s_{b,60}$$
$$= 461 \ ft - 400.3 \ ft$$
$$= 60.7 \ ft$$

Determine the perception-reaction time required at 60 mph.

$$t_{r,60} = \frac{s_{r,60}}{v}$$

$$= \left(\frac{60.7 \ ft}{60 \ \frac{mi}{hr}}\right)\left(\frac{3600 \ \frac{sec}{hr}}{5280 \ \frac{ft}{mi}}\right)$$

$$= 0.690 \ sec \quad (0.7 \ sec)$$

The answer is (C).

Why Other Options Are Wrong

(A) This incorrect answer results from an improper conversion from miles per hour to feet per second.

(B) This incorrect answer results from an improper conversion from miles per hour to feet per second, ignoring the negative value and assuming the answer must be a positive value.

(D) This incorrect answer results from neglecting the perception-reaction distance at 50 mph to establish the braking friction factor, thereby assuming an average friction factor for the entire stopping distance. The answer assumes a positive value.

SOLUTION 8

The lead car stops in 278 ft, which is the braking distance, $s_{b,lead}$. The braking distance determines the minimum friction factor, f.

$$s_{b,lead} = \frac{v_1^2 - v_2^2}{2g(f+G)}$$

Set v_2 at 0 mph, and solve for f.

$$f = \frac{v_1^2}{2gs_{b,50}}$$

$$= \frac{\left(\left(50 \ \frac{mi}{hr}\right)\left(5280 \ \frac{ft}{mi}\right)\right)^2}{(2)\left(32.2 \ \frac{ft}{sec^2}\right)(278 \ ft)\left(3600 \ \frac{sec}{hr}\right)^2}$$

$$= 0.30$$

The stopping distance, s_s, for the following car includes braking distance, s_b, and perception-reaction distance, s_r.

$$s_s = s_r + s_b$$

Find the braking distance from 60 mph using the friction factor for the lead car.

$$s_{b,60} = \frac{v_1^2 - v_2^2}{2g(f+G)}$$

$$= \left(\frac{\left(60 \ \frac{mi}{hr}\right)^2 - \left(0 \ \frac{mi}{hr}\right)}{(2)\left(32.2 \ \frac{ft}{sec^2}\right)(0.30)}\right)\left(\frac{5280 \ \frac{ft}{mi}}{3600 \ \frac{sec}{hr}}\right)^2$$

$$= 400.3 \ ft$$

Determine the perception-reaction distance for the following car.

$$s_{r,60} = vt_r$$

$$= \frac{\left(60 \ \frac{mi}{hr}\right)\left(5280 \ \frac{ft}{mi}\right)(2.0 \ sec)}{3600 \ \frac{sec}{hr}}$$

$$= 176 \ ft$$

Determine the stopping distance from 60 mph, $s_{s,60}$.

$$s_{s,60} = s_{r,60} + s_{b,60}$$
$$= 176 \text{ ft} + 400.3 \text{ ft}$$
$$= 576.3 \text{ ft}$$

Determine the minimum following distance.

$$s_f = s_{s,60} - s_{b,50}$$
$$= 576.3 \text{ ft} - 278 \text{ ft}$$
$$= 298.3 \text{ ft} \quad (300 \text{ ft})$$

The answer is (C).

Why Other Options Are Wrong

(A) This incorrect answer results from assuming that the braking distance for the lead car includes perception-reaction time (2.0 sec).

(B) This answer results from incorrectly subtracting the length of the lead car (20 ft standard from *A Policy on Geometric Design of Highways and Streets* (GDHS)) from the available stopping distance for the following car.

(D) This incorrect answer results from adding the length of the lead car (20 ft standard from GDHS) to the stopping distance available for the following car.

SOLUTION 9

The total distance between stations has three components of travel.

$$s_{\text{total}} = s_{\text{accel}} + s_{\text{running}} + s_{\text{decel}} = 5280 \text{ ft}$$

Determine the distance necessary to accelerate to 80 mph and the distance necessary to decelerate from 80 mph to a stop.

$$s_{\text{accel}} = \frac{v_2^2 - v_1^2}{2a}$$
$$= \left(\frac{\left(80 \ \frac{\text{mi}}{\text{hr}} \right)^2 - \left(0 \ \frac{\text{mi}}{\text{hr}} \right)^2}{(2)\left(5.5 \ \frac{\text{ft}}{\text{sec}^2} \right)\left(3600 \ \frac{\text{sec}}{\text{hr}} \right)^2} \right) \left(5280 \ \frac{\text{ft}}{\text{mi}} \right)^2$$
$$= 1252 \text{ ft}$$

$$s_{\text{decel}} = \frac{v_2^2 - v_1^2}{2d}$$
$$= \left(\frac{\left(0 \ \frac{\text{mi}}{\text{hr}} \right)^2 - \left(80 \ \frac{\text{mi}}{\text{hr}} \right)^2}{(2)\left(-4.4 \ \frac{\text{ft}}{\text{sec}^2} \right)\left(3600 \ \frac{\text{sec}}{\text{hr}} \right)^2} \right) \left(5280 \ \frac{\text{ft}}{\text{mi}} \right)^2$$
$$= 1564 \text{ ft}$$

Subtract the acceleration and deceleration distances from 1 mi to find the distance of constant running at 80 mph.

$$s_{\text{running}} = 5280 \text{ ft} - s_{\text{accel}} - s_{\text{decel}}$$
$$= 5280 \text{ ft} - 1252 \text{ ft} - 1564 \text{ ft}$$
$$= 2464 \text{ ft}$$

Determine the constant running speed time.

$$t_{\text{running}} = \frac{s_{\text{running}}}{v}$$
$$= \left(\frac{2464 \text{ ft}}{\left(80 \ \frac{\text{mi}}{\text{hr}} \right)\left(5280 \ \frac{\text{ft}}{\text{mi}} \right)} \right) \left(3600 \ \frac{\text{sec}}{\text{hr}} \right)$$
$$= 21.0 \text{ sec}$$

Determine the acceleration time.

$$t_{\text{accel}} = \frac{v_2 - v_1}{a}$$
$$= \frac{\left(80 \ \frac{\text{mi}}{\text{hr}} \right)\left(5280 \ \frac{\text{ft}}{\text{mi}} \right) - 0 \ \frac{\text{mi}}{\text{hr}}}{\left(5.5 \ \frac{\text{ft}}{\text{sec}^2} \right)\left(3600 \ \frac{\text{sec}}{\text{hr}} \right)}$$
$$= 21.3 \text{ sec}$$

Determine the deceleration time.

$$t_{\text{decel}} = \frac{v_2 - v_1}{d}$$
$$= \frac{0 \ \frac{\text{mi}}{\text{hr}} - \left(80 \ \frac{\text{mi}}{\text{hr}} \right)\left(5280 \ \frac{\text{ft}}{\text{mi}} \right)}{\left(-4.4 \ \frac{\text{ft}}{\text{sec}^2} \right)\left(3600 \ \frac{\text{sec}}{\text{hr}} \right)}$$
$$= 26.7 \text{ sec}$$

The total time to travel between stations is the sum of the component times.

$$t_{\text{total}} = t_{\text{accel}} + t_{\text{running}} + t_{\text{decel}}$$
$$= 21.3 \text{ sec} + 21.0 \text{ sec} + 26.7 \text{ sec}$$
$$= 69.0 \text{ sec}$$

Determine the average travel speed between stations.

$$v = \frac{s}{t} = \left(\frac{1 \text{ mi}}{69.0 \text{ sec}}\right)\left(3600 \, \frac{\text{sec}}{\text{hr}}\right)$$
$$= 52.2 \text{ mph} \quad (52 \text{ mph})$$

The answer is (C).

Why Other Options Are Wrong

(A) Using the 5 min train schedule spacing as the time required to travel from one station to the next gives a result that is too small.

(B) This incorrect answer results from using 80 mph as the top instant speed, ignoring the 80 mph constant running time, and proportioning the acceleration and deceleration rates to determine the acceleration and deceleration times covering the entire 1 mi distance between stations.

(D) This incorrect answer results from failing to convert from miles per hour to feet per second.

SOLUTION 10

Determine the total number of train cars arriving per hour.

$$\text{total trains per hour} = \frac{1}{\text{train spacing}}$$
$$= \left(\frac{1 \text{ train}}{5 \text{ min}}\right)\left(60 \, \frac{\text{min}}{\text{hr}}\right)$$
$$= 12 \text{ trains/hr}$$

$$\text{no. of cars per hour} = (\text{no. of trains per hour})$$
$$\times (\text{no. of cars per train})$$
$$= \left(12 \, \frac{\text{trains}}{\text{hr}}\right)\left(4 \, \frac{\text{cars}}{\text{train}}\right)$$
$$= 48 \text{ cars/hr}$$

Multiply the number of cars per hour by the capacity of each car to determine the number of passengers per hour.

$$\left(48 \, \frac{\text{cars}}{\text{hr}}\right)\left(220 \, \frac{\text{passengers}}{\text{car}}\right) = 10{,}560 \text{ passengers/hr}$$
$$(10{,}600 \text{ passengers/hr})$$

The answer is (C).

Why Other Options Are Wrong

(A) This incorrect answer results from not including the number of trains in each hour.

(B) This incorrect answer results from not including the number of cars in each train.

(D) This incorrect answer results from subtracting the dwell time from the arrival time, which is the time between trains, and then using the time between trains to determine the number of trains per hour.

SOLUTION 11

The total distance between stations includes acceleration distance and deceleration distance only.

$$s_{\text{total}} = s_{\text{accel}} + s_{\text{decel}}$$

Determine the acceleration distance to 150 mph using the given acceleration rate.

$$s_{\text{accel}} = \frac{v_2^2 - v_1^2}{2a}$$
$$= \frac{\left(\left(150 \, \frac{\text{mi}}{\text{hr}}\right)\left(5280 \, \frac{\text{ft}}{\text{mi}}\right)\right)^2 - \left(0 \, \frac{\text{mi}}{\text{hr}}\right)^2}{(2)(0.18)\left(32.2 \, \frac{\text{ft}}{\text{sec}^2}\right)\left(3600 \, \frac{\text{sec}}{\text{hr}}\right)^2}$$
$$= 4175 \text{ ft}$$

Determine the deceleration distance from 150 mph using the given deceleration rate.

$$s_{\text{decel}} = \frac{v_1^2 - v_2^2}{2d}$$
$$= \frac{\left(0 \, \frac{\text{mi}}{\text{hr}}\right)^2 - \left(\left(150 \, \frac{\text{mi}}{\text{hr}}\right)\left(5280 \, \frac{\text{ft}}{\text{mi}}\right)\right)^2}{(2)(-0.12)\left(32.2 \, \frac{\text{ft}}{\text{sec}^2}\right)\left(3600 \, \frac{\text{sec}}{\text{hr}}\right)^2}$$
$$= 6263 \text{ ft}$$

Determine the total distance to accelerate to 150 mph then decelerate to a stop.

$$s_{\text{total}} = \dfrac{4175 \text{ ft} + 6263 \text{ ft}}{5280 \dfrac{\text{ft}}{\text{mi}}}$$

$$= 1.98 \text{ mi} \quad (2.0 \text{ mi})$$

The minimum station spacing is approximately 2.0 mi.

The answer is (C).

Why Other Options Are Wrong

(A) This incorrect distance only covers acceleration to 150 mph.

(B) This answer results from an incorrect conversion from miles per hour to feet per second.

(D) This incorrect answer results from using the metric value of gravity acceleration without converting to feet per second.

SOLUTION 12

Saturation flow rate is determined from *Highway Capacity Manual* (HCM) Eq. 19-8.

$$s = s_o f_w f_{\text{HV}g} f_p f_{\text{bb}} f_a f_{\text{LU}} f_{\text{LT}} f_{\text{RT}} f_{Lpb} f_{Rpb} f_{\text{wz}} f_{\text{ms}} f_{\text{sp}}$$
$$s_o = 1900 \text{ pcphpl}$$
$$f_w = 1.00 \quad [\text{HCM Exh. 19-20}]$$

Solve for the following.

$$f_{\text{HV}g} = \dfrac{100 - 0.3 P_{\text{HV}} - 2.07 P_g}{100}$$
$$= \dfrac{100 - 0 - (2.07)(4)}{100}$$
$$= 0.92$$

$$f_p = \dfrac{N - 0.1 - \dfrac{18 N_m}{3600}}{N}$$
$$= \dfrac{1 - 0.1 - \dfrac{(18)(20)}{3600}}{1}$$
$$= 0.80$$

$$f_{\text{bb}} = \dfrac{N - \dfrac{14.4 N_B}{3600}}{N}$$
$$= \dfrac{1 - \dfrac{(14.4)(0)}{3600}}{1}$$
$$= 1.00$$

$$f_a = 0.90 \quad [\text{for CBD}]$$

$$f_{\text{LU}} = 1.0 \quad [\text{HCM p. 19-47}]$$

$$f_{\text{LT}} = \dfrac{1}{1.0 + 0.05 P_{\text{LT}}}$$
$$= \dfrac{1}{1.0 + (0.05)(0)}$$
$$= 1.00$$

Use HCM Eq. 19-13 to adjust for the right-turn shared lane turning path. Apply $E_R = 1.18$ only to the proportion of right-turning vehicles.

$$f_{\text{RT}} = 1 - \left(\dfrac{1}{E_R}\right)\left(\dfrac{V_{\text{RT}}}{V}\right)$$
$$= 1 - \left(\dfrac{1}{1.18}\right)\left(\dfrac{139 \dfrac{\text{veh}}{\text{hr}}}{139 \dfrac{\text{veh}}{\text{hr}} + 731 \dfrac{\text{veh}}{\text{hr}}}\right)$$
$$= 0.86$$

$$f_{Lpb} = 1.00$$

$$f_{Rpb} = 1.00$$

Determine the saturation flow rate.

$$s = \left(1900 \dfrac{\text{pc}}{\text{hr-ln}}\right)(1.00)(1.00)(0.92)(0.80)(1.00)(0.90)$$
$$\times (1.00)(1.00)(0.86)(1.00)(1.00)(1.00)$$
$$= 1082 \text{ vphpl}$$

There is only one lane, so the total saturation flow rate is 1082 vph (1080 vph).

The answer is (B).

Why Other Options Are Wrong

(A) This incorrect answer is the sum of the intersection movements for approach E, or the approach volume.

(C) This incorrect answer is the result of neglecting to consider the approach grade.

(D) This incorrect answer is the result of considering the area type factor as 1.0 for CBD instead of 0.90.

SOLUTION 13

The number of buses required is determined by dividing the number of bus riders (passengers) by the maximum capacity of each bus.

$$\text{no. of buses} = \frac{\text{no. of bus riders}}{\text{capacity of each bus}}$$

The peak hour factor is the peak 15 min flow rate compared to the peak 1 hr flow rate. The arena will empty in 1 hour, so the peak 1 hr flow is 12,000 persons. Determine how many people will ride buses, V, during one hour following the event.

$$
\begin{aligned}
V &= (\text{fraction of bus riders})(\text{people in attendance}) \\
&= (0.35)(12{,}000 \text{ persons}) \\
&= 4200 \text{ bus riders}
\end{aligned}
$$

Define the peak hour factor (PHF), where V_{15} is the flow rate in persons per hour during the peak 15 min period.

$$
\begin{aligned}
\text{PHF} &= \frac{\text{flow volume in peak 1 hr}}{4(\text{flow volume in peak 15 min})} \\
&= \frac{V}{4V_{15}}
\end{aligned}
$$

Rearrange the definition to determine the number of bus riders during the 15 min peak.

$$
\begin{aligned}
V_{15} &= \frac{V}{4(\text{PHF})} \\
&= \frac{4200 \text{ bus riders}}{(4)(0.85)} \\
&= 1235 \text{ bus riders}
\end{aligned}
$$

Determine the number of buses needed during the peak 15 min period.

$$
\begin{aligned}
\frac{\text{bus riders}}{\text{bus capacity}} &= \frac{1235 \text{ bus riders}}{62 \dfrac{\text{bus riders}}{\text{bus}}} \\
&= 19.9 \text{ buses} \quad (20 \text{ buses})
\end{aligned}
$$

The answer is (B).

Why Other Options Are Wrong

(A) This incorrect answer assumes that the peak hour factor is applied to the total hour and then divided into 15 min periods.

(C) This incorrect answer projects the peak number of buses to an hourly demand.

(D) This incorrect answer divides the bus patronage into buses and then misapplies the peak hour factor to the resulting number of buses.

SOLUTION 14

A magnetic compass can show the heading under no power and poor visibility conditions. A universal default numbering system is a fail-safe feature of aviation, so that pilots can land planes with the greatest degree of safety. Without a unified numbering system, confusion would result, especially among international flights.

The answer is (A).

Why Other Options Are Wrong

(B) When taking off, the pilot already knows the plane's direction. The need to find the correct landing heading under flight conditions is vastly more critical. The greater necessity, then, would favor landing headings.

(C) Assigning random numbers by various designers would lead to massive confusion between airports. Leaving the choice to designers could lead to the adoption of many systems. That fact, and the problem of pride of authorship, would interfere with the adoption of a universal system.

(D) Using bearings requires additional letters, N-S-E-W, and the result would be repeated numbers. Clarity of radio communications would not be good with different pronunciations of letters in different languages.

SOLUTION 15

The effective width based on a unit flow rate for a required density of flow is determined by *Highway Capacity Manual* (HCM) Eq. 24-3.

$$ v_p = \frac{v_{15}}{15 W_E} $$

Determine the number of persons leaving in the first hour.

$$
\begin{aligned}
\text{persons leaving} &= (\text{person capacity}) \\
\text{in first hour} \quad &\times (\text{fraction leaving in first hour}) \\
&= (50{,}000 \text{ persons})(0.95) \\
&= 47{,}500 \text{ persons}
\end{aligned}
$$

Determine the transit volume.

$$
\begin{aligned}
\text{transit volume} &= (\text{persons leaving in first hour}) \\
&\times (\text{fraction transit riders}) \\
&= (47{,}500 \text{ persons})(0.35) \\
&= 16{,}625 \text{ transit riders}
\end{aligned}
$$

Define the peak hour factor (PHF), where V_{15} is the flow rate in persons per hour during the peak 15 min period.

$$\text{PHF} = \frac{V}{4 V_{15}}$$

Determine the peak 15 min volume.

$$V_{15} = \frac{V}{4(\text{PHF})} = \frac{16{,}625 \text{ transit riders}}{(4)(0.75)}$$

$$= 5542 \text{ transit riders}$$

Find the flow rate from HCM Exh. 24-2. The minimum width for LOS D utilizes the maximum flow rate of 11 peds/ft-min.

Determine the effective width required using HCM Eq. 24-3.

$$v_p = \frac{v_{15}}{15 \, W_E} = \frac{5542 \text{ persons}}{(15 \text{ min})\left(11 \, \dfrac{\text{persons}}{\text{ft-min}}\right)}$$

$$= 33.6 \text{ ft} \quad (34 \text{ ft})$$

The answer is (B).

Why Other Options Are Wrong

(A) This incorrect answer uses the flow rate for LOS D, unadjusted for platoons, as described in HCM Exh. 24-1.

(C) This incorrect answer results from a misapplication of the PHF.

(D) This incorrect answer results from determining the width needed for the entire stadium exit in the peak hour.

SOLUTION 16

Determine the number of cars arriving in the design hour.

$$V = (\text{fraction of vehicles arriving in design hour})$$
$$\times (\text{lot capacity})$$
$$= \left(\frac{0.72}{\text{hr}}\right)(700 \text{ veh})$$
$$= 504 \text{ vph}$$

Compare the required n for each PHF value.

$$n_1 = \frac{504 \, \dfrac{\text{veh}}{\text{hr}}}{(0.70)\left(270 \, \dfrac{\text{veh}}{\text{hr}}\right)} = 2.7$$

$$n_2 = \frac{504 \, \dfrac{\text{veh}}{\text{hr}}}{(0.80)\left(270 \, \dfrac{\text{veh}}{\text{hr}}\right)} = 2.3$$

$$n_3 = \frac{504 \, \dfrac{\text{veh}}{\text{hr}}}{(0.87)\left(270 \, \dfrac{\text{veh}}{\text{hr}}\right)} = 2.1$$

$$n_4 = \frac{504 \, \dfrac{\text{veh}}{\text{hr}}}{(0.94)\left(270 \, \dfrac{\text{veh}}{\text{hr}}\right)} = 2.0$$

$$n_5 = \frac{504 \, \dfrac{\text{veh}}{\text{hr}}}{(0.97)\left(270 \, \dfrac{\text{veh}}{\text{hr}}\right)} = 1.9$$

Summarize.

no. of lanes	PHF	$(\text{PHF})\mu$ (vph)	n_{required} (lanes)
1	0.70	189	2.7
2	0.80	216	2.3
3	0.87	235	2.1
4	0.94	254	2.0
5 or more	0.97	265	1.9

The equation balances at three lanes.

The answer is (B).

Why Other Options Are Wrong

(A) This incorrect answer does not adjust the service rate by the peak hour factor.

(C) This answer incorrectly uses the total lot capacity as the incoming volume during the design hour.

(D) This answer results from incorrectly assuming that the peak hour factor is the actual flow proportion during the peak 15 min period instead of the flow *rate* during the peak 15 min period.

Traffic Engineering

SOLUTION 17

Use the base relationship.

$$D = \frac{\text{unit length}}{\text{bicycle spacing}}$$

Determine the following.

$$V = (\text{directional factor})(\text{two-way flow})$$
$$= (0.6)\left(\frac{20 \text{ bicycles}}{15 \text{ min}}\right)\left(60 \frac{\text{min}}{\text{hr}}\right)$$
$$= 48 \text{ bicycles/hr}$$

The time between bicycle arrivals is

$$h = \frac{\text{time unit}}{\text{arrivals during time unit}}$$
$$= \frac{60 \frac{\text{min}}{\text{hr}}}{48 \frac{\text{bicycles}}{\text{hr}}}$$
$$= 1.25 \text{ min/bicycle}$$

$$S = \frac{\left(11.2 \frac{\text{mi}}{\text{hr}}\right)\left(5280 \frac{\text{ft}}{\text{mi}}\right)}{3600 \frac{\text{sec}}{\text{hr}}}$$
$$= 16.4 \text{ ft/sec}$$

$$\text{bicycle spacing} = hS$$
$$= \left(1.25 \frac{\text{min}}{\text{bicycle}}\right)\left(16.4 \frac{\text{ft}}{\text{sec}}\right)\left(60 \frac{\text{sec}}{\text{min}}\right)$$
$$= 1230 \text{ ft/bicycle}$$

$$D = \frac{5280 \frac{\text{ft}}{\text{mi}}}{1230 \frac{\text{ft}}{\text{bicycle}}}$$
$$= 4.29 \text{ bicycles/mi} \quad (4.3 \text{ bicycles/mi})$$

The answer is (B).

Alternate Solution

$$D = \frac{V}{S} = \frac{48 \frac{\text{bicycles}}{\text{hr}}}{11.2 \frac{\text{mi}}{\text{hr}}}$$
$$= 4.3 \text{ bicycles/mi}$$

Why Other Options Are Wrong

(A) This incorrect answer results from inverting the density determination.

(C) This incorrect answer results from using the two-directional flow as the one-way flow.

(D) This incorrect answer results from incorrectly applying the density and flow rate relationship.

SOLUTION 18

Travel between stations includes three conditions of motion: acceleration, travel at constant speed, and deceleration. Travel at constant speed uses the distance remaining between acceleration and deceleration distances.

$$s_{\text{total}} = s_{\text{accel}} + s_{\text{running}} + s_{\text{decel}}$$

Determine the acceleration distance.

$$s_{\text{accel}} = \frac{v_2^2 - v_1^2}{2a}$$
$$= \frac{\left(\left(80 \frac{\text{mi}}{\text{hr}}\right)\left(5280 \frac{\text{ft}}{\text{mi}}\right)\right)^2 - \left(0 \frac{\text{mi}}{\text{hr}}\right)^2}{(2)\left(5.5 \frac{\text{ft}}{\text{sec}^2}\right)\left(3600 \frac{\text{sec}}{\text{hr}}\right)^2}$$
$$= 1252 \text{ ft}$$

Determine the deceleration distance.

$$s_{\text{decel}} = \frac{v_2^2 - v_1^2}{2d}$$
$$= \frac{\left(0 \frac{\text{mi}}{\text{hr}}\right)^2 - \left(\left(80 \frac{\text{mi}}{\text{hr}}\right)\left(5280 \frac{\text{ft}}{\text{mi}}\right)\right)^2}{(2)\left(-4.5 \frac{\text{ft}}{\text{sec}^2}\right)\left(3600 \frac{\text{sec}}{\text{hr}}\right)^2}$$
$$= 1530 \text{ ft}$$

Find the distance available for constant running speed.

$$s_{\text{running}} = s_{\text{total}} - s_{\text{accel}} - s_{\text{decel}}$$
$$= (1 \text{ mi})\left(5280 \frac{\text{ft}}{\text{mi}}\right) - 1252 \text{ ft} - 1530 \text{ ft}$$
$$= 2498 \text{ ft}$$

The total travel time is the sum of acceleration time, constant-speed running time, and deceleration time.

$$t_{\text{total}} = t_{\text{accel}} + t_{\text{running}} + t_{\text{decel}}$$

Determine the acceleration time.

$$t_{accel} = \frac{v_2 - v_1}{a}$$

$$= \frac{\left(80 \ \frac{mi}{hr}\right)\left(5280 \ \frac{ft}{mi}\right) - 0 \ \frac{mi}{hr}}{\left(5.5 \ \frac{ft}{sec^2}\right)\left(3600 \ \frac{sec}{hr}\right)}$$

$$= 21.3 \ sec$$

Determine the constant running speed time.

$$t_{running} = \frac{s_{running}}{v}$$

$$= \left(\frac{2498 \ ft}{80 \ \frac{mi}{hr}}\right)\left(\frac{3600 \ \frac{sec}{hr}}{5280 \ \frac{ft}{mi}}\right)$$

$$= 21.3 \ sec$$

Determine the deceleration time.

$$t_{decel} = \frac{v_2 - v_1}{d}$$

$$= \frac{0 \ \frac{mi}{hr} - \left(80 \ \frac{mi}{hr}\right)\left(5280 \ \frac{ft}{mi}\right)}{\left(-4.5 \ \frac{ft}{sec^2}\right)\left(3600 \ \frac{sec}{hr}\right)}$$

$$= 26.1 \ sec$$

Determine the total time.

$$t_{total} = 21.3 \ sec + 21.3 \ sec + 26.1 \ sec$$

$$= 68.7 \ sec$$

The average running speed is the sum of the distances traveled divided by the sum of the times traveled. Determine the average running speed.

$$v_{ave} = \frac{s_{total}}{t_{total}}$$

$$= \left(\frac{5280 \ ft}{68.7 \ sec}\right)\left(\frac{3600 \ \frac{sec}{hr}}{5280 \ \frac{ft}{mi}}\right)$$

$$= 52.4 \ mph \quad (52 \ mph)$$

The answer is (C).

Why Other Options Are Wrong

(A) This answer results from incorrectly including the dwell time at both stations in the total time between two stations.

(B) This answer results from incorrectly including the dwell time at one station in the total time between stations.

(D) This answer results from not converting miles per hour to feet per second.

SOLUTION 19

The average speed is the total running time plus the delay per stop divided into the 10 mi length of trip.

$$v_{ave} = \frac{s_{total}}{t_{running} + (30 \ sec)(\text{number of stops})}$$

Determine the running time without stopping.

$$t = \frac{s_{total}}{v_{running}}$$

$$= \left(\frac{10 \ mi}{35 \ \frac{mi}{hr}}\right)\left(60 \ \frac{min}{hr}\right)$$

$$= 17.1 \ min$$

Add the delay time for the current number of stops.

$$t_{delay} = (\text{no. of stops per mile})(\text{length of line})$$
$$\times (\text{delay per stop})$$

$$= \frac{\left(4 \ \frac{stops}{mi}\right)(10 \ mi)\left(30 \ \frac{sec}{stop}\right)}{60 \ \frac{sec}{min}}$$

$$= 20 \ min$$

The current time for one trip is the running time plus the stop delay time.

$$t_{total} = t_{running} + t_{delay}$$
$$= 17.1 \ min + 20 \ min$$
$$= 37.1 \ min$$

Determine the current average speed.

$$v_{ave} = \frac{s_{total}}{t_{total}}$$

$$= \left(\frac{10 \ mi}{37.1 \ min}\right)\left(60 \ \frac{min}{hr}\right)$$

$$= 16.2 \ mph$$

Determine the delay time for the proposed number of stops.

$$t_{\text{new delay}} = \frac{\left(3 \, \frac{\text{stops}}{\text{mi}}\right)(10 \text{ mi})\left(30 \, \frac{\text{sec}}{\text{stop}}\right)}{60 \, \frac{\text{sec}}{\text{min}}}$$
$$= 15 \text{ min}$$

Determine the proposed trip time.

$$t_{\text{prop}} = t_{\text{running}} + t_{\text{new delay}}$$
$$= 17.1 \text{ min} + 15 \text{ min}$$
$$= 32.1 \text{ min}$$

Determine the proposed average speed.

$$v_{\text{prop}} = \frac{s}{t_{\text{prop}}}$$
$$= \left(\frac{10 \text{ mi}}{32.1 \text{ min}}\right)\left(60 \, \frac{\text{min}}{\text{hr}}\right)$$
$$= 18.7 \text{ mph}$$

Determine the change in average speed.

$$\Delta v_{\text{ave}} = v_{\text{prop}} - v_{\text{ave}}$$
$$= 18.7 \, \frac{\text{mi}}{\text{hr}} - 16.2 \, \frac{\text{mi}}{\text{hr}}$$
$$= 2.5 \text{ mph}$$

The answer is (B).

Why Other Options Are Wrong

(A) This incorrect answer results from eliminating one deceleration and one acceleration from each mile, but not from the dwell time.

(C) This incorrect answer is the difference between the no-stop speed of 35 mph and the proposed speed for 3 stops/mi.

(D) This incorrect answer is the proposed average speed.

SOLUTION 20

The service flow rate is the hourly equivalent flow rate that occurs in the peak 15 min.

Passenger-car equivalents are determined by the heavy vehicle factor, criteria II, and the driver population factor, criteria I, if known to have an effect on capacity.

The number of lanes included in the vehicle count, criteria IV, must be known so that the count can be averaged to a single lane.

The peak-hour factor, criteria VI, is used to determine how much of the peak-hour traffic flows in the peak 15 min.

The answer is (B).

Why Other Options Are Wrong

(A) The volume-over-capacity ratio, criteria VIII, is determined after the service flow rate is determined and is not a direct factor in determining service flow rate.

(C) Lane width and parking adjustment factors, criteria III and V, affect service volume determinations for intersection approaches, not service flow rates for highways.

(D) Lane width and parking adjustment factors, criteria III and V, affect service volume determinations for intersection approaches, not service flow rates for highways. Average control delay per vehicle, criteria VII, applies to controlled intersections.

SOLUTION 21

HCM Exh. 12-26 shows that as the percentage of trucks increases beyond 2%, the passenger-car equivalence decreases, particularly on longer grades.

The answer is (B).

Why Other Options Are Wrong

(A) Trucks generally operate much slower than other traffic on steep downgrades because of stopping safety. Trucks therefore have a higher passenger-car equivalent on steep downgrades than they do on slight downgrades.

(C) For general freeway sections, the most common traffic pattern exhibits a morning peak flow and an evening peak flow, which is attributable to auto commuters. Truck traffic, on the other hand, tends to increase during mid-day near urban areas because of local delivery activity. Therefore, trucks usually make up a lower percentage of peak-hour traffic than of off-peak traffic.

(D) From HCM Exh. 3-15, passenger car equivalents are assigned to single-unit trucks (FHWA class 4–7), and tractor-trailer combinations (FHWA class 8–13), which includes all trucks with two or more units, one of which is a straight truck or tractor power unit. Single-unit trucks are not divided into light and heavy groups.

SOLUTION 22

For general traffic analysis, the HCM uses standard intervals of either one-hour or 15-minutes. The 15-minute intervals are multiplied by four to determine one-hour flow rates. The peak-hour factor is calculated by dividing a peak one-hour flow rate by the peak 15-minute flow rate within that hour.

The answer is (D).

Why Other Options Are Wrong

(A) The peak 15-minute flow rate divided by the peak one-hour flow rate yields a simple flow fraction, but not a comparison of flow rates.

(B) While other analysis methods, such as peak within a peak, may use a five-minute flow rate, the HCM uses a 15-minute flow rate for the peak-hour factor.

(C) The peak five-minute flow rate divided by the peak one-hour flow rate yields a simple flow fraction, but not a comparison of flow rates.

SOLUTION 23

Trucks descending long, steep downgrades generally travel more slowly than they do on long upgrades, in order to avoid loss of braking power and loss of directional control. Slower speeds on severe downgrades equate to larger passenger-car equivalents than on upgrades.

The answer is (C).

Why Other Options Are Wrong

(A) Buses in a traffic stream increase the passenger capacity far beyond the decrease in vehicle capacity based on auto equivalents.

(B) Trucks require more auto-equivalent space and usually carry no more, if not fewer, people than autos. Therefore, the passenger capacity of the roadway is lower.

(D) According to the *Highway Capacity Manual* (HCM), RVs in small proportion to trucks can be included with trucks.

SOLUTION 24

Determine the UTF.

$$\text{UTF} = \left(\frac{1}{1000}\right)\left(\frac{\text{households}}{\text{auto}}\right)\left(\frac{\text{persons}}{\text{mi}^2}\right)$$

$$= \left(\frac{1}{1000}\right)\left(\frac{1\ \text{household}}{0.80\ \text{auto}}\right)\left(10{,}000\ \frac{\text{persons}}{\text{mi}^2}\right)$$

$$= 12.50$$

Determine the percentage of trips on transit.

$$\%\ \text{transit usage} = \frac{\text{UTF}}{0.6} = \frac{12.50}{0.6}$$

$$= 20.8\%$$

Determine the number of residents expected to use transit.

$$\begin{aligned} \text{no. of residents} \\ \text{using transit} &= (\text{transit usage})(\text{no. of trips}) \\ &= (0.208)(12{,}000) \\ &= 2500\ \text{persons} \end{aligned}$$

The answer is (D).

Why Other Options Are Wrong

(A) This incorrect answer results from applying the UTF directly to the population to obtain transit users.

(B) This incorrect answer results from reversing the auto density to 0.80 households per auto instead of 0.80 autos per household.

(C) This incorrect answer results from applying the transit percentage to the population density instead of to the total trips.

SOLUTION 25

Determine the total round trip vehicle-miles traveled.

$$\text{VM}_{\text{total}} = s\left[P_c P_{\text{total}}\left(\frac{1}{O_c}\right) + P_{\text{sov}} P_{\text{total}}\left(\frac{1}{O_{\text{sov}}}\right)\right]$$

$$= (2)(8\ \text{mi})\left(\begin{array}{l} (0.25)(7000\ \text{pers})\left(\dfrac{1}{2.10\ \frac{\text{pers}}{\text{veh}}}\right) \\[3ex] + (0.75)(7000\ \text{pers})\left(\dfrac{1}{1\ \frac{\text{pers}}{\text{veh}}}\right) \end{array} \right)$$

$$= 97{,}333\ \text{veh-mi}$$

Determine the fuel consumption rate per vehicle.

$$F = 0.0362\ \frac{\text{gal}}{\text{veh-mi}} + \left(\frac{0.746\ \frac{\text{gal}}{\text{veh-hr}}}{20\ \frac{\text{mi}}{\text{hr}}}\right)$$

$$= 0.0735\ \text{gal/veh-mi}$$

Determine the gallons of fuel consumed.

$$\begin{aligned} \text{total fuel consumption} &= (\text{total vehicle-miles})F \\ &= (97{,}333\ \text{veh-mi}) \\ &\quad \times\left(0.0735\ \frac{\text{gal}}{\text{veh-mi}}\right) \\ &= 7154\ \text{gal} \end{aligned}$$

Determine the Btu use.

$$
\begin{aligned}
\text{Btu use} &= (\text{total fuel consumption})\left(\frac{\text{Btu content}}{\text{gal}}\right) \\
&= (7154 \text{ gal})\left(125{,}000 \ \frac{\text{Btu}}{\text{gal}}\right) \\
&= 8.94 \times 10^8 \text{ Btu} \quad (8.9 \times 10^8 \text{ Btu})
\end{aligned}
$$

The answer is (C).

Why Other Options Are Wrong

(A) This is the number of gallons consumed per day.

(B) This is the total one-way Btu consumption.

(D) This is the Btu consumption if all 7000 commuters arrived in a single-occupant vehicle.

SOLUTION 26

Improvement to public safety in urban areas involves providing an adequate physical environment so that individuals do not feel insecure or uncertain about what is happening nearby. All of the listed items can improve the feeling of personal comfort or reduce the level of discomfort with surroundings. The traveling public is surrounded by noise and vibration in an urban setting at a fairly constant level. Conceivably, the noise level in a transit terminal, option D, would not be much different from that on the street. The other three options could be perceived as making a greater difference within the transit terminal than in the surrounding neighborhood, thereby improving the safety of the terminal itself.

The answer is (D).

Why Other Options Are Wrong

(A) Improving personal security is an inherent goal of improving public safety.

(B) Reliability in transit service reduces anxiety about unknown arrival times. This reduces the need for longer waiting times in the terminal, especially during off-peak travel times.

(C) Adequate lighting levels significantly improve the travelers' confidence in their surroundings and lead to an improved feeling of personal safety.

SOLUTION 27

Determine the number of person-trips per household.

$$
\begin{aligned}
T &= 0.78 + 1.3P + 2.3A \\
&= 0.78 + (1.3)\left(3.5 \ \frac{\text{persons}}{\text{household}}\right) \\
&\quad + (2.3)\left(2.2 \ \frac{\text{autos}}{\text{household}}\right) \\
&= 10.39 \text{ trips/household-day}
\end{aligned}
$$

Determine the number of person-trips in the entire zone.

$$
\begin{aligned}
N &= T(\text{no. of households}) \\
&= \left(10.39 \ \frac{\text{trips}}{\text{household-day}}\right)(600 \text{ households}) \\
&= 6234 \text{ trips/day}
\end{aligned}
$$

Determine the number of auto trips per day.

$$
\begin{aligned}
T_A &= (\text{fraction of auto trips})\left(\frac{\text{total trips}}{\text{day}}\right) \\
&= (0.94)\left(6234 \ \frac{\text{trips}}{\text{day}}\right) \\
&= 5860 \text{ trips/day} \quad (5900 \text{ trips/day})
\end{aligned}
$$

The answer is (B).

Why Other Options Are Wrong

(A) This is the number of trips per household.

(C) This is the total number of trips per day for the zone, including transit and other modes.

(D) This answer results from misuse of the modal split ratio.

SOLUTION 28

The hourly vehicle volume is a count of the general mix of vehicle types, usually reported in units of vehicles per hour. The design hourly flow rate must reflect the influence of heavy vehicles, the hourly variation of traffic, and the characteristics of the driver population. The design equivalent passenger-car flow rate is calculated using the heavy-vehicle and peak-hour adjustment factors and includes the driver population adjustment.

The answer is (B).

Why Other Options Are Wrong

(A) This incorrect answer is an inaccurate definition of peak hour factor.

(C) This incorrect answer does not convert to passenger-car equivalents.

(D) This incorrect answer does not correct for passenger-car equivalents.

SOLUTION 29

The directional capacity is the capacity per lane times the number of lanes in each direction.

$$\text{directional capacity} = \left(\frac{\text{capacity}}{\text{hr-lane}}\right)(\text{no. of lanes})$$

$$= \left(2100\ \frac{\text{veh}}{\text{hr-lane}}\right)(2\ \text{lanes})$$

$$= 4200\ \text{vph}$$

The arrival rate is more than the one-lane capacity. Therefore, when one lane is blocked, there will be a queue of vehicles behind the blockage. Once the blocked lane is opened, the capacity will be greater than the arrival rate, so the queue will begin to dissipate. The total queue to be dissipated is as follows.

$$\begin{aligned}
\text{total queue} \atop \text{length} &= \sum \text{arrivals} - \sum \text{departures} \\
&= \text{arrivals during blockage} \\
&\quad + \text{arrivals during queue dissipation} \\
&\quad - \text{departures during lane blockage} \\
&\quad - \text{departures during queue dissipation}
\end{aligned}$$

Arrivals for 20 min plus arrivals for time t equals departures at the one-lane rate for 20 min plus departures at the two-lane rate for time t.

Determine the arrival rate.

$$\text{arrival rate} = \frac{3100\ \dfrac{\text{veh}}{\text{hr}}}{60\ \dfrac{\text{min}}{\text{hr}}}$$

$$= 51.7\ \text{vpm}$$

Determine the two-lane departure rate, which is the full directional capacity.

$$\text{two-lane departure rate} = \frac{4200\ \dfrac{\text{veh}}{\text{hr}}}{60\ \dfrac{\text{min}}{\text{hr}}}$$

$$= 70\ \text{vpm}$$

Determine the one-lane departure rate, which is the departure rate during the incident blockage (see *Highway Capacity Manual* (HCM) Exh. 11-23).

$$\text{incident blockage departure rate} = \frac{\left(70\ \dfrac{\text{veh}}{\text{min}}\right)(0.70)}{2}$$

$$= 24.5\ \text{vpm}$$

Departure time, t, is the time necessary to dissipate the queue. The time is 0 sec at the instant of clearing the blockage. The total queue length is

$$\begin{aligned}
\text{total} \atop \text{queue length} &= \left(51.7\ \frac{\text{veh}}{\text{min}}\right)(20\ \text{min}) + \left(51.7\ \frac{\text{veh}}{\text{min}}\right)t \\
&= \left(24.5\ \frac{\text{veh}}{\text{min}}\right)(20\ \text{min}) + \left(70\ \frac{\text{veh}}{\text{min}}\right)t \\
t &= 29.7\ \text{min} \quad (30\ \text{min})
\end{aligned}$$

The entire queue will be dissipated in approximately 30 min.

The answer is (B).

Why Other Options Are Wrong

(A) This incorrect answer results from including in the calculation only the vehicle(s) that arrive during the blockage.

(C) This incorrect answer results from adding the 20 min blockage time to the queue dissipation time after the blockage.

(D) This incorrect answer results from not subtracting the single-lane departures during the 20 min lane blockage.

SOLUTION 30

The skid distance is

$$s_b = \frac{v_1^2 - v_2^2}{2g(f + G)}$$

$$= \frac{(v_1^2 - v_2^2)\left(5280 \dfrac{ft}{mi}\right)^2}{(2)\left(32.2 \dfrac{ft}{sec^2}\right)\left(3600 \dfrac{sec}{hr}\right)^2 (f + G)}$$

$$= \frac{v_1^2 - v_2^2}{\left(30 \dfrac{mi^2}{hr^2\text{-}ft}\right)(f + G)}$$

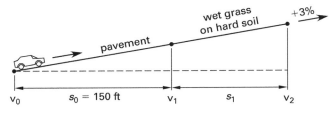

Solve for the initial speed approaching the skid on wet grass.

$$v_1 = \sqrt{v_0^2 + \left(-30(f + G)\right)s_0}$$

$$= \sqrt{\begin{array}{c}\left(60 \dfrac{mi}{hr}\right)^2 + \left(\left(-30 \dfrac{mi^2}{hr^2\text{-}ft}\right)\left(0.30 + 0.03 \dfrac{ft}{ft}\right)\right) \\ \times (150 \text{ ft})\end{array}}$$

$$= 46 \text{ mph}$$

Determine the skid distance on the grass.

$$s_1 = \frac{v_2^2 - v_1^2}{-30(f + G)}$$

$$= \frac{\left(0 \dfrac{mi}{hr}\right)^2 - \left(46 \dfrac{mi}{hr}\right)^2}{\left(-30 \dfrac{mi^2}{hr^2\text{-}ft}\right)\left(0.10 + 0.03 \dfrac{ft}{ft}\right)}$$

$$= 542 \text{ ft} \quad (540 \text{ ft})$$

The answer is (A).

Why Other Options Are Wrong

(B) This incorrect answer results from not including the effect of the grade on the stopping distance.

(C) This incorrect answer results from not using a negative conversion constant, −30, to denote deceleration.

(D) This incorrect answer results from inserting the speed value in feet per second in place of the speed value in miles per hour.

SOLUTION 31

The braking distance is

$$s_b = \frac{v_1^2 - v_2^2}{2g(f + G)}$$

$$= \frac{(v_1^2 - v_2^2)\left(5280 \dfrac{ft}{mi}\right)^2}{(2)\left(32.2 \dfrac{ft}{sec^2}\right)\left(3600 \dfrac{sec}{hr}\right)^2 (f + G)}$$

$$= \frac{v_1^2 - v_2^2}{\left(30 \dfrac{mi^2}{hr^2\text{-}ft}\right)(f + G)}$$

The grade is negative since the car skids downhill. This has the effect of reducing the initial speed for a given length of the skid.

Rearrange the equation and solve for v_2.

$$v_2 = \sqrt{v_1^2 - s_b 30(f + G)}$$

$$= \sqrt{\begin{array}{c}\left(70 \dfrac{mi}{hr}\right)^2 - (350 \text{ ft})\left(30 \dfrac{mi^2}{hr^2\text{-}ft}\right) \\ \times \left(0.30 - 0.05 \dfrac{ft}{ft}\right)\end{array}}$$

$$= 47.7 \text{ mph} \quad (48 \text{ mph})$$

The answer is (C).

Why Other Options Are Wrong

(A) This incorrect answer results from adding the grade to the friction factor instead of subtracting.

(B) This incorrect answer results from not including the grade but considering tire friction only.

(D) This incorrect answer results from assuming v_2 is 0 mph and reducing the form to the equation to solve for v.

SOLUTION 32

The final speed is 0 mph. The stopping distance is

$$s_b = \frac{v_1^2}{2g(f+g)}$$

$$= \frac{v_1^2 \left(5280 \dfrac{\text{ft}}{\text{mi}}\right)^2}{(2)\left(32.2 \dfrac{\text{ft}}{\text{sec}^2}\right)\left(3600 \dfrac{\text{sec}}{\text{hr}}\right)^2 (f+G)}$$

$$= \frac{v_1^2}{\left(30 \dfrac{\text{mi}^2}{\text{hr}^2\text{-ft}}\right)(f+G)}$$

Solve for the total braking distance from 60 mph assuming the car did not strike the concrete barriers.

$$s_b = \frac{\left(60 \dfrac{\text{mi}}{\text{hr}}\right)^2}{\left(30 \dfrac{\text{mi}^2}{\text{hr}^2\text{-ft}}\right)\left(0.30 + 0.03 \dfrac{\text{ft}}{\text{ft}}\right)}$$

$$= 364 \text{ ft}$$

The car skidded 150 ft before the impact. Solve for the additional distance to stop if the concrete barriers had not been encountered.

$$D_b = s_b - \text{skid distance before impact}$$
$$= 364 \text{ ft} - 150 \text{ ft}$$
$$= 214 \text{ ft} \quad (210 \text{ ft})$$

The answer is (B).

Why Other Options Are Wrong

(A) This incorrect answer results from improper conversion between miles per hour and feet per second.

(C) This incorrect answer results from ignoring the grade.

(D) This incorrect answer results from subtracting the grade from the friction factor instead of adding it to the friction factor.

SOLUTION 33

The braking distance is

$$s_b = \frac{v_2^2 - v_1^2}{2d}$$

Rearrange for the initial speed.

$$v_1 = \sqrt{v_2^2 - 2ds_b}$$

Determine the acceleration (or, in this case, deceleration) rate using the friction factor and the grade.

$$d = g(f+G) = \left(9.81 \dfrac{\text{m}}{\text{s}^2}\right)\left(0.48 + \left(-0.06 \dfrac{\text{m}}{\text{m}}\right)\right)$$
$$= 4.12 \text{ m/s}^2$$

The acceleration is negative.

$$v_1 = \sqrt{\frac{\left(\left(40 \dfrac{\text{km}}{\text{h}}\right)\left(1000 \dfrac{\text{m}}{\text{km}}\right)\right)^2}{\left(3600 \dfrac{\text{s}}{\text{h}}\right)^2} - (2)\left(-4.12 \dfrac{\text{m}}{\text{s}^2}\right)(60 \text{ m})}$$

$$= 24.9 \text{ m/s}$$

Convert to kilometers per hour.

$$v_1 = \frac{\left(24.9 \dfrac{\text{m}}{\text{s}}\right)\left(3600 \dfrac{\text{s}}{\text{h}}\right)}{1000 \dfrac{\text{m}}{\text{km}}}$$

$$= 89.5 \text{ kph} \quad (90 \text{ kph})$$

The answer is (C).

Why Other Options Are Wrong

(A) This incorrect answer is the initial speed in miles per second, not kilometers per hour.

(B) This incorrect answer results from not considering acceleration as negative and then ignoring the negative sign under the square root.

(D) This incorrect answer results from not including the effect of grade on deceleration.

SOLUTION 34

The deceleration equation has two unknowns, the initial velocity, v_1, and the friction factor, f, which can be solved by trial and error.

$$s = \frac{v_1^2}{2g(f+G)} = \frac{v_1^2}{(2)\left(32.2 \dfrac{\text{ft}}{\text{sec}^2}\right)(f+G)}$$

Assume the initial speed is 40 mph. From the table, the friction factor is 0.51.

$$v_1 = \sqrt{s(2g)(f+G)}$$
$$= \sqrt{(185 \text{ ft})\left((2)\left(32.2 \frac{\text{ft}}{\text{sec}^2}\right)\right)\left(0.51 - 0.03 \frac{\text{ft}}{\text{ft}}\right)}$$
$$= 75.6 \text{ ft/sec}$$

Convert to miles per hour.

$$v_1 = \frac{\left(75.6 \frac{\text{ft}}{\text{sec}}\right)\left(3600 \frac{\text{sec}}{\text{hr}}\right)}{5280 \frac{\text{ft}}{\text{mi}}}$$
$$= 51.6 \text{ mph}$$

This is greater than 40 mph.

Try an initial speed of 50 mph. The friction factor is 0.45.

$$v_1 = \sqrt{(185 \text{ ft})\left((2)\left(32.2 \frac{\text{ft}}{\text{sec}^2}\right)\right)\left(0.45 - 0.03 \frac{\text{ft}}{\text{ft}}\right)}$$
$$= 70.7 \text{ ft/sec}$$

Convert to miles per hour.

$$v_1 = \frac{\left(70.7 \frac{\text{ft}}{\text{sec}}\right)\left(3600 \frac{\text{sec}}{\text{hr}}\right)}{5280 \frac{\text{ft}}{\text{mi}}}$$
$$= 48.2 \text{ mph}$$

This result is less than 50 mph. Therefore, the initial speed must be between 40 mph and 50 mph. Try an initial speed of 49 mph.

Determine the friction factor by linear interpolation.

$$f = \frac{0.51 - 0.45}{10} + 0.45$$
$$= 0.456$$

Use this value to determine the initial speed.

$$v_1 = \sqrt{(185 \text{ ft})\left((2)\left(32.2 \frac{\text{ft}}{\text{sec}^2}\right)\right)\left(0.456 - 0.03 \frac{\text{ft}}{\text{ft}}\right)}$$
$$= 71.24 \text{ ft/sec}$$

Convert to miles per hour.

$$v_1 = \frac{\left(71.24 \frac{\text{ft}}{\text{sec}}\right)\left(3600 \frac{\text{sec}}{\text{hr}}\right)}{5280 \frac{\text{ft}}{\text{mi}}}$$
$$= 48.6 \text{ mph} \quad (49 \text{ mph})$$

The answer is (B).

Why Other Options Are Wrong

(A) This incorrect answer results from improper rationalizing of the equation throughout the trial-and-error process.

(C) This incorrect answer results from ignoring the downgrade of the skid.

(D) This incorrect answer is the initial speed in feet per second, not miles per hour.

SOLUTION 35

For a segment of highway, the crash rates are reported per HMVM traveled per year using the following formula. Injury crashes include both fatal and nonfatal injuries.

$$R = \frac{(\text{no. of injury crashes})(10^8)}{(\text{ADT})(\text{no. of sample years})\left(365 \frac{\text{days}}{\text{yr}}\right)L}$$

Determine the number of injury crashes.

$$\begin{aligned}\text{injury crashes} &= \text{total fatal crashes} \\ &\quad + \text{total personal injury crashes} \\ &= (0+3+2+5) + (35+35+50+50) \\ &= 180 \text{ crashes}\end{aligned}$$

Determine the ADT.

$$\begin{aligned}\text{ADT}_{\text{ave}} &= \frac{\sum \text{ADT}}{\text{no. of sample years}} \\ &= \frac{\begin{array}{c}15{,}500 \frac{\text{veh}}{\text{day}} + 16{,}000 \frac{\text{veh}}{\text{day}} \\ + 16{,}500 \frac{\text{veh}}{\text{day}} + 17{,}000 \frac{\text{veh}}{\text{day}}\end{array}}{4} \\ &= 16{,}250 \text{ vpd}\end{aligned}$$

Determine the injury crash rate.

$$R = \frac{(180 \text{ crashes})\left(10^8 \dfrac{\text{veh-mi}}{\text{HMVM}}\right)}{\left(16{,}250 \dfrac{\text{veh}}{\text{day}}\right)(4 \text{ yr})\left(365 \dfrac{\text{days}}{\text{yr}}\right)(20 \text{ mi})}$$

$$= 37.93 \text{ crashes/HMVM} \quad (38 \text{ crashes/HMVM})$$

The answer is (C).

Why Other Options Are Wrong

(A) This is the injury rate per million annual vehicle miles instead of per hundred million annual vehicle miles.

(B) This incorrect answer results from erroneously counting only the personal injury crashes in the rate, neglecting to include the fatal crashes.

(D) This incorrect answer results from erroneously counting all crashes in the rate, including property damage only.

SOLUTION 36

The total queue to be dissipated is the sum of the 10 min vehicle accumulation plus the arrivals that continue while the queue is dissipating. The departure rate from the front of the queue is at the roadway capacity.

Determine the arrival rate.

$$\text{arrival rate} = \frac{3100 \dfrac{\text{veh}}{\text{hr}}}{60 \dfrac{\text{min}}{\text{hr}}}$$

$$= 51.7 \text{ vpm}$$

Determine the departure rate.

$$\text{departure rate} = \frac{4200 \dfrac{\text{veh}}{\text{hr}}}{60 \dfrac{\text{min}}{\text{hr}}}$$

$$= 70 \text{ vpm}$$

Equate the vehicle accumulation to the departures.

$$\left(51.7 \frac{\text{veh}}{\text{min}}\right)(10 \text{ min})$$

$$+\left(51.7 \frac{\text{veh}}{\text{min}}\right)t = \left(70 \frac{\text{veh}}{\text{min}}\right)t$$

$$t = 28.2 \text{ min} \quad (28 \text{ min})$$

This queue will be dissipated approximately 28 min after the freeway is reopened to full capacity.

The answer is (B).

Why Other Options Are Wrong

(A) This incorrect answer results from only including the vehicles that arrive during the blockage in the calculation.

(C) This incorrect answer results from including the 10 min blockage time in the departure time.

(D) This incorrect answer results from subtracting the 10 min blockage time from the departure time.

SOLUTION 37

While all of the countermeasures listed could lead to crash injury reduction, only ergonomic design of vehicle interiors is not directly involved with highway design, construction, operation, or maintenance. This is covered by other agencies and industry standards.

The answer is (B).

Why Other Options Are Wrong

(A) Pedestrians use roads, streets, and highways and must be accommodated in a safe fashion. Pedestrian management is a function of street and highway management responsibility.

(C) Enforcement of traffic laws, in conjunction with signing and marking programs, is a function of highway traffic management responsibility.

(D) Signing, marking, and delineation are an integral part of on-highway safety improvements.

SOLUTION 38

Slippery road conditions require drivers to allow more spacing between vehicles than is normal for a given speed. Reducing the approach speed limit would encourage drivers to compensate for the poor surface, with little effect on traffic volume. Visibility improvements would naturally occur by having traffic approach the intersection at a lower speed, which in turn would allow more time for reacting to unusual conditions at the intersection.

The answer is (D).

Why Other Options Are Wrong

(A) Improved roadway lighting would possibly improve visibility at night, but would have a negligible effect on conflicts caused by reduced traction.

(B) Overhead signals may or may not be warranted. Unless there is already a serious restriction on visibility of signals by approaching traffic, additional signals would probably have a negligible effect on conflicts caused by reduced traction.

(C) Prohibiting turns may eliminate some crashes, depending on traffic volumes and movement patterns. There would be a negligible effect on conflicts caused by reduced traction.

SOLUTION 39

Left-turning vehicles require sharing of space with oncoming traffic flow. Large straight-through flow volumes combined with large left-turn flow volumes compete for time in the conflict zone shared by both flows.

Adequate visibility aided by roadway lighting is important, but is of little advantage if there are inadequate gaps in the flow to allow left-turning vehicles to cross the opposing flow.

If the intersection is signalized, a special left-turning phase can create needed gaps in traffic flow and would be more important than additional roadway lighting in most cases.

The answer is (B).

Why Other Options Are Wrong

(A) Inadequate gaps in the opposing flow of traffic encourage left-turn drivers to take more risks.

(C) The absence of a left-turning phase would generally be more important than the need for additional roadway lighting.

(D) A large volume of left turns is the most likely cause of head-on collisions.

SOLUTION 40

Deficiencies that are most related to efficient use of road space are I, II, III, V, VI, and VIII.

Deficiencies I, III, and VI relate to capacity and level of service, which are measures of roadway efficiency.

Deficiencies II and V relate to managing traffic flow.

Deficiencies IV and VII relate to construction standards that could have an effect on traffic flow under severe conditions but that are not as directly related as other deficiencies listed.

Deficiency IX relates to the number of truck companies servicing businesses. The number of companies has no bearing on traffic flow; rather, it is the number of trucks that is important.

The deficiencies least likely to qualify for TSM programs are IV, drainage; VII, lighting; and IX, the number of truck companies.

The answer is (B).

Why Other Options Are Wrong

(A) Deficiency III, the peak average travel speed below 20 mph, is strongly related to traffic management and efficient use of road space.

(C) Deficiencies VI, low level of service, and VIII, high crash rate, strongly relate to traffic management and efficient use of road space.

(D) Deficiency V, sight distance limitations, strongly relates to traffic management and efficient use of road space.

SOLUTION 41

Determine the annual traffic base.

$$\text{traffic base} = (\text{ADT})\left(365 \ \frac{\text{days}}{\text{yr}}\right)L$$

$$= \frac{\left(28{,}000 \ \dfrac{\text{veh}}{\text{day}}\right)\left(365 \ \dfrac{\text{days}}{\text{yr}}\right)(2.5 \ \text{mi})}{10^8 \ \dfrac{\text{HMVM}}{\text{veh-mi}}}$$

$$= 0.256 \ \text{HMVM/yr}$$

Determine the critical rate for the statewide average crash rate.

$$R_{\text{crit}} = R_{\text{ave}} + K\sqrt{\frac{R_{\text{ave}}}{\text{traffic base}}}$$

$$= 150 \ \frac{\text{crashes}}{\text{HMVM}} + 1.645\sqrt{\frac{150 \ \dfrac{\text{crashes}}{\text{HMVM}}}{0.256 \ \text{HMVM}}}$$

$$= 190 \ \text{crashes/HMVM}$$

Determine the crash rate for the particular highway segment being evaluated.

$$R_{\text{seg}} = \frac{\text{annual crash rate}}{\text{traffic base}}$$

$$= \frac{9 \ \dfrac{\text{crashes}}{\text{yr}}}{0.256 \ \dfrac{\text{HMVM}}{\text{yr}}}$$

$$= 35.2 \ \text{crashes/HMVM}$$

Determine the ratio of the segment crash rate with respect to the statewide critical rate.

$$\frac{R_{\text{seg}}}{R_{\text{crit}}} = \frac{35.2 \dfrac{\text{crashes}}{\text{HMVM}}}{190 \dfrac{\text{crashes}}{\text{HMVM}}}$$
$$= 0.185 \quad (0.19)$$

The answer is (B).

Why Other Options Are Wrong

(A) This incorrect answer results from applying the test factor directly to the statewide crash rate.

(C) This answer results from incorrectly determining the critical rate.

(D) This incorrect answer results from comparing the statewide critical rate to the segment crash rate.

SOLUTION 42

Find the base total number of roadway segment crashes per year using *Highway Safety Manual* (HSM) Eq. 11-9 and Table 11-5.

$$N_{\text{spfrd}} = e^{(a + b(\ln \text{AADT}) + \ln L)}$$
$$= e^{(-9.025 + (1.049)(\ln 18,000) + \ln 2)}$$
$$= 7.003 \text{ crashes/yr}$$

Check against the graph of SPF shown in HSM Fig. 11-4.

For an AADT of 18,000 veh/day, the approximate average crash frequency is ± 3.5 crash/mi. For a 2 mi segment, $2 \pm 3.5 = \pm 7.0$ crash/mi. Therefore, the calculated base total number of crashes for this segment is correct.

Adjust the base total to reflect the site-specific geometric conditions and determine the crash modification factors, CMF_{1rd} through CMF_{5rd}, for divided roadway segments. From HSM Table 11-16, the CMF for a 12 ft lane width is 1.00, the default value for the proportion of total crashes constituted by related crashes, p_{ra}, is 0.50, and CMF_{ra} is 1.03.

$$\text{CMF}_{\text{1rd}} = p_{\text{ra}}(\text{CMF}_{\text{ra}} - 1.0) + 1.0$$
$$= (0.50)(1.03 - 1.0) + 1.0$$
$$= 1.015$$

The CMF for a 6 ft wide paved right shoulder is determined from HSM Table 11-17.

$$\text{CMF}_{\text{2rd}} = 1.04$$

The CMF for an 8 ft wide median without a barrier is approximated from HSM Table 11-18.

$$\text{CMF}_{\text{3rd}} = 1.04$$

The CMF for lighting is

$$\text{CMF}_{\text{4rd}} = 1.00 \quad \text{[base condition]}$$

The CMF for speed enforcement is

$$\text{CMF}_{\text{5rd}} = 1.00 \quad \text{[base condition]}$$

Using HSM Eq. 11-3, the predicted average crash frequency for the road segment is

$$N_{\text{predicted rs}} = N_{\text{spfrd}} C_r \text{CMF}_{\text{1rd}} \text{CMF}_{\text{2rd}}$$
$$\times \text{CMF}_{\text{3rd}} \text{CMF}_{\text{4rd}} \text{CMF}_{\text{5rd}}$$
$$= \left(7.003 \frac{\text{crashes}}{\text{yr}}\right)(0.94)(1.015)(1.04)$$
$$\times (1.04)(1.00)(1.00)$$
$$= 7.227 \text{ crashes/yr} \quad (7.23 \text{ crashes/yr})$$

The answer is (C).

Why Other Answers Are Wrong

(A) This incorrect answer results from using a 1 mi long segment, instead of 2 mi.

(B) This incorrect answer results from using a 1 mi long segment, instead of 2 mi, and not applying the local calibration factor.

(D) This incorrect answer results from not applying the local calibration factor.

SOLUTION 43

Using HSM Eq. 10-8, the predicted average crash frequency of a three-leg, stop-controlled intersection is

$$N_{\text{spf3ST}} = \exp\left(\begin{array}{c} -9.86 + 0.79 \ln(\text{AADT}_{\text{maj}}) \\ +0.49 \ln(\text{AADT}_{\text{min}}) \end{array}\right)$$
$$= \exp\left(\begin{array}{c} -9.86 + 0.79 \ln\left(7000 \dfrac{\text{veh}}{\text{day}}\right) \\ +0.49 \ln\left(1000 \dfrac{\text{veh}}{\text{day}}\right) \end{array}\right)$$
$$= 1.681 \text{ crashes/yr}$$

Determine the crash modification factors, CMF_{1i} through CMF_{4i}, for the intersection. The CMF for a $0°$ intersection skew angle is

$$CMF_{1i} = 1.0 \quad [0° \text{ skew from } 90°]$$

The CMF for left-turn lanes is

$$CMF_{2i} = 1.0 \quad \begin{bmatrix} \text{no left-turn lanes} \\ \text{base condition} \end{bmatrix}$$

The CMF for right-turn lanes is

$$CMF_{3i} = 1.0 \quad \begin{bmatrix} \text{no right-turn lanes} \\ \text{base condition} \end{bmatrix}$$

The CMF for lighting is

$$CMF_{4i} = 1.0 \quad \begin{bmatrix} \text{absence of intersection} \\ \text{lighting base condition} \end{bmatrix}$$

The combined CMF is

$$
\begin{aligned}
CMF_{comb} &= (CMF_{1i})(CMF_{2i})(CMF_{3i})(CMF_{4i}) \\
&= (1.0)(1.0)(1.0)(1.0) \\
&= 1.0
\end{aligned}
$$

Crash data within 5% of the statewide mean indicates the calibration factor, C_i, can be assumed to be 1.05 to account for the maximum predicted crash rate.

Using HSM Eq. 10-3, the predicted average crash frequency for the intersection is

$$
\begin{aligned}
N_{\text{predicted int}} &= N_{\text{spf int}} C_i CMF_{comb} \\
&= \left(1.681 \ \frac{\text{crashes}}{\text{yr}}\right)(1.05)(1.0) \\
&= 1.765 \ \text{crashes/yr} \quad (1.77 \ \text{crashes/yr})
\end{aligned}
$$

The answer is (D).

Why Other Answers Are Wrong

(A) This incorrect answer does not use the exponential function to determine $N_{\text{spf 3ST}}$.

(B) This incorrect answer applies the left-turn CMF for a three-leg intersection, equal to 0.56.

(C) This incorrect answer does not apply the local calibration factor.

2 Horizontal Design

PROBLEM 1

A circular horizontal curve with a radius of 1430 ft and a deflection angle of 14°30′ Rt has its point of intersection (PI) at sta 572+00. The forward tangent is to be shifted parallel to itself and 12 ft to the left.

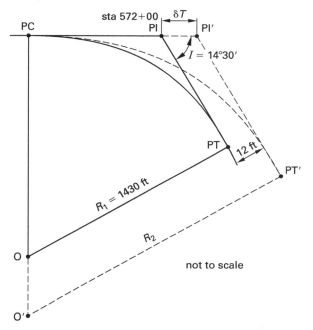

not to scale

If the point of curvature (PC) station remains in the same place, what is most nearly the radius of the shifted curve?

(A) 1050 ft

(B) 1530 ft

(C) 1810 ft

(D) 1820 ft

Hint: The curve deflection remains the same for the shifted curve. A new tangent length is increased by extending the back tangent to intersect the shifted position of the ahead tangent.

PROBLEM 2

In the following illustration, subtangent AB is 330 ft, angle α is 25°, and angle β is 38°.

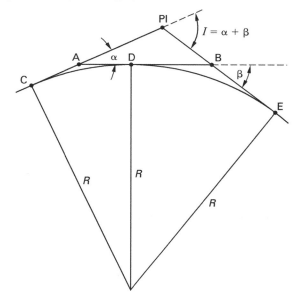

What is most nearly the radius of the curve that will be tangent to lines CI, AB, and IE?

(A) 270 ft

(B) 320 ft

(C) 580 ft

(D) 4300 ft

Hint: The angles given are the deflections of each of the curve arcs, which can be treated as two independent curves with the same radius.

PROBLEM 3

In the following illustration, AG is 200 ft, angle α is 90°, and the degree of curve is 2°.

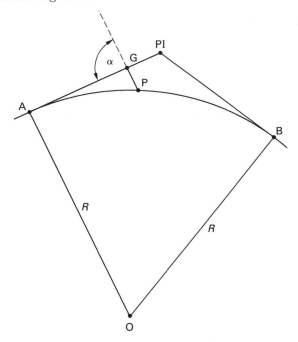

What is most nearly the distance GP?

(A)   1.8 ft

(B)   3.5 ft

(C)   7.0 ft

(D)   28 ft

Hint: Construct a triangle using point P as one corner.

PROBLEM 4

A horizontal curve has a deflection angle of 67°45′ at its point of intersection (PI), which is at sta 126+50.

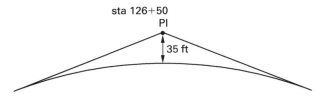

When the external distance is close to 35 ft, what is most nearly the radius of the curve to the nearest whole degree of curve?

(A)   44 ft

(B)   170 ft

(C)   180 ft

(D)   210 ft

Hint: Solving initially for the radius, then converting to degree of curve reduces the number of steps and the chance of errors.

PROBLEM 5

A circular horizontal curve has a radius of 2500 ft and a deflection angle of 35°. The point of intersection (PI) of the curve is at sta 485+26.75.

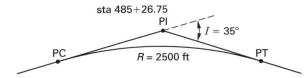

What is the station of the point of tangent (PT)?

(A)   sta 483+00

(B)   sta 492+40

(C)   sta 492+65

(D)   sta 493+15

Hint: First, find the station of the point of curvature (PC) based on the tangent length of the curve.

PROBLEM 6

A curve on a low-speed urban street has a superelevation of −4%. The design is to follow AASHTO's *A Policy on Geometric Design of Highways and Streets* (GDHS) for a design speed of 30 mph. What is most nearly the minimum radius for the curve?

(A)   16 ft

(B)   98 ft

(C)   250 ft

(D)   375 ft

Hint: A reverse cross slope increases the feeling of discomfort when driving a curve at a particular speed.

PROBLEM 7

What is most nearly the north coordinate of the point of tangent (PT) shown?

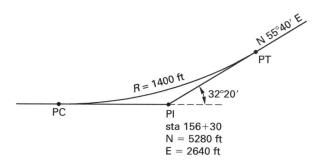

(A) 2975 ft

(B) 5500 ft

(C) 5510 ft

(D) 5615 ft

Hint: Find the tangent distance using the radius and curve deflection. The bearing of the back tangent is not needed to solve the problem.

PROBLEM 8

A curve to the right of a ramp has a radius of 200 ft with the point of curvature (PC) of the curve at sta 10+00. A pier corner at sta 10+50 is offset 33.33 ft to the right of the curve's tangent.

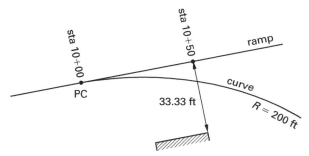

Most nearly, what is the radial clearance from the ramp curve to the pier corner?

(A) 26 ft

(B) 27 ft

(C) 34 ft

(D) 41 ft

Hint: Set up a triangle using a line from the center of the curve through the pier corner, extended to the point on the curve that is radial to the pier corner.

PROBLEM 9

An obstructed curve is shown. Stopping sight distances for various design speeds are given.

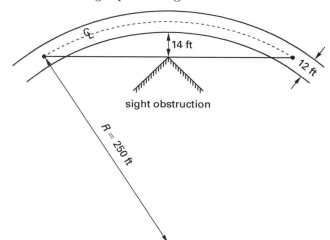

design speed (mph)	stopping sight distance (ft)
15	75
25	150
30	200
35	250

What is the maximum design speed for the curve?

(A) 15 mph

(B) 25 mph

(C) 30 mph

(D) 35 mph

Hint: The obstruction offset distance is the mid-ordinate of the circular curve.

PROBLEM 10

A specification for highway superelevation transition stipulates that the change in cross slope cannot exceed 0.02 ft/ft for each second of travel. A right-hand curve is being designed with full superelevation of 0.08 ft/ft on a four-lane divided freeway. The lanes are 12 ft wide, and the design speed is 70 mph. The normal slope is 0.01 ft/ft down to the right. What is most nearly the minimum length of superelevation transition for the right side of the road?

(A) 100 ft

(B) 360 ft

(C) 410 ft

(D) 460 ft

Hint: The distance traveled in one second is the change in cross slope (superelevation) of 0.02 ft/ft.

PROBLEM 11

A 6° railroad curve is to be paralleled by a highway centerline, which is to be 125 ft to the inside of the railroad curve. The railroad is laid out using chord definition curves, while the highway is to use arc definition curves. What is the degree of curve for the highway?

(A) 5° 18′ 12″

(B) 6° 54′ 00″

(C) 6° 54′ 13″

(D) 16° 12′ 56″

Hint: Solve the problem using radius definition, then convert back to degree of curve.

PROBLEM 12

For the curve shown, what is the bearing of the ahead tangent?

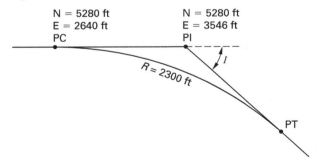

(A) 133°

(B) S 43° E

(C) S 47° E

(D) S 68° 30′ E

Hint: The curve deflection can be determined using the curve radius and curve tangent length.

PROBLEM 13

A specification for highway superelevation transition stipulates that the edge profile transition is not to exceed a 1:200 difference from the centerline profile. A right-hand curve is being designed with full superelevation of 0.08 ft/ft on a four-lane divided freeway. The lanes are 12 ft wide, and the design speed is 70 mph. The normal slope is 0.01 ft/ft, and the superelevation is to be rotated about the median edge of the running lanes. What is most nearly the minimum length of transition for the right side of the roadway?

(A) 200 ft

(B) 340 ft

(C) 390 ft

(D) 2800 ft

Hint: Determine how much elevation difference will occur in the right edge of the pavement.

PROBLEM 14

Superelevation transition for a 150 mph railway is determined by the rotational change of the vehicle on U.S. standard gage track. Specifications call for an American Railway Engineering and Maintenance-of-Way Association (AREMA) design of 1.15° cross slope change per second at the design speed. What is most nearly the transition length for a 5 in superelevation?

(A) 250 ft

(B) 660 ft

(C) 970 ft

(D) 11,000 ft

Hint: Use a standard railroad gage of 56.5 in.

PROBLEM 15

An engineer is laying out a highway centerline that has a fully spiraled curve at its point of intersection (PI), which is at sta 196+00. The curve deflection is 24° Rt, and the central curve has a radius of 850 ft. Using a 250 ft spiral, what is most nearly the external distance of the curve?

(A) 19 ft

(B) 22 ft

(C) 32 ft

(D) 84 ft

Hint: The central curve of a fully spiraled curve is shifted inward to account for the offset effect of the spiral.

PROBLEM 16

A six-lane divided highway is being designed with a spiraled horizontal 5.73° curve and a 60 mph running speed. The longer of two criteria is to be used to set the length of spiral on the inside curve lanes. The first criterion is

$$L_{s,ft} = \left(1.6 \; \frac{ft^2\text{-}hr^3}{mi^3}\right)\left(\frac{v_{mph}^3}{R_{ft}}\right)$$

The other criterion is that the pavement edge profile should not deviate more than 1:200 from the centerline profile in the transition, using the spiral length as the superelevation transition. The normal slope of the roadway is 0.015 ft/ft, and the maximum superelevation is to be 0.08 ft/ft. Superelevation is to rotate about the median edge, and the lanes are 12 ft wide. Most nearly, what length of spiral should be used on the inside curve lanes?

(A)   350 ft

(B)   470 ft

(C)   580 ft

(D)   950 ft

Hint: Find the change in elevation of the outside edge of the pavement in relation to the centerline.

PROBLEM 17

A compound spiral transition curve is to connect two compound horizontal circular curves. The design speed is 70 mph. The length of spiral transition is to be determined by

$$L_{s,ft} = \left(1.6 \; \frac{ft^2\text{-}hr^3}{mi^3}\right)\left(\frac{v_{mph}^3}{R_{ft}}\right)$$

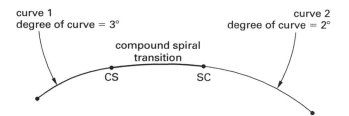

What is the approximate length of spiral required?

(A)   100 ft

(B)   190 ft

(C)   290 ft

(D)   480 ft

Hint: The change in radius from one curve to the next is related to time.

PROBLEM 18

A reverse spiral transition is to connect two reverse horizontal circular curves on a roadway. The design speed is 60 mph. The length of spiral transition is to be determined by

$$L_{s,ft} = \left(1.6 \; \frac{ft^2\text{-}hr^3}{mi^3}\right)\left(\frac{v_{mi/hr}^3}{R_{ft}}\right)$$

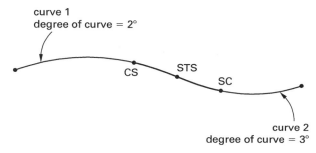

Most nearly, what is the total required length of spiral from the curve-to-spiral point (CS) to the spiral-to-curve point (SC)?

(A)   60 ft

(B)   120 ft

(C)   180 ft

(D)   300 ft

Hint: A spiral carries the degree of curve to zero, then from zero to the new degree of curve in the opposite direction.

Horizontal
Design

PROBLEM 19

An existing simple horizontal curve is to have spirals added to the ends. The center portion of the curve is to remain unshifted. The degree of curve is 5°, the design speed is 50 mph, and the length of spiral transition is 200 ft.

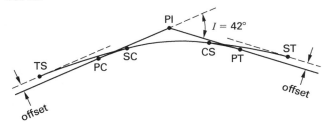

Approximately how many feet should the tangents be shifted in order to accommodate the spirals?

(A)   −14.2 ft

(B)   1.46 ft

(C)   2.55 ft

(D)   5.82 ft

Hint: The spiral offset is the same as p for a spiral.

SOLUTION 1

Find the PC station of the unshifted curve. First, determine the curve tangent length.

$$T = R_1 \tan \frac{I}{2}$$
$$= (1430 \text{ ft})\left(\tan \frac{14°30'}{2}\right)$$
$$= 181.92 \text{ ft}$$

Subtract the curve tangent from the PI station to find the PC station.

PI sta 572	+ 00
− 1	81.92
PC sta 570	+ 18.08

Find the length of the new tangent. Let δT be the added length to extend to the shifted position of the ahead tangent.

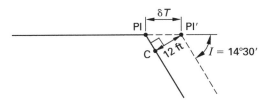

not to scale

$$\delta T = \frac{\text{offset}}{\sin I} = \frac{12.00 \text{ ft}}{\sin 14°30'}$$
$$= 47.93 \text{ ft}$$
$$\text{new } T = T + \delta T = 181.92 \text{ ft} + 47.93 \text{ ft}$$
$$= 229.85 \text{ ft}$$
$$R_2 = \frac{\text{new } T}{\tan \frac{I}{2}} = \frac{229.85 \text{ ft}}{\tan \frac{14°30'}{2}}$$
$$= 1806.74 \text{ ft} \quad (1810 \text{ ft})$$

The answer is (C).

Why Other Options Are Wrong

(A) This incorrect answer is the radius if the ahead tangent is shifted to the right.

(B) This incorrect answer is the radius if cosine (instead of sine) is used to find the tangent extension.

(D) This incorrect answer is the radius if the deflection angle is entered into the calculation as 14.3° instead of 14.5°.

SOLUTION 2

The sum of the two curve arc deflections, α and β, equals the total curve deflection.

Let C, D, and E be the points of tangency.

$$CA = AD = R \tan \frac{\alpha}{2}$$

$$DB = BE = R \tan \frac{\beta}{2}$$

Combine.

$$
\begin{aligned}
AB &= AD + DB \\
&= R\left(\tan \frac{\alpha}{2} + \tan \frac{\beta}{2}\right) \\
&= 330 \text{ ft}
\end{aligned}
$$

Rearrange.

$$
\begin{aligned}
R &= \frac{AB}{\tan \dfrac{\alpha}{2} + \tan \dfrac{\beta}{2}} \\
&= \frac{330 \text{ ft}}{\tan \dfrac{25°}{2} + \tan \dfrac{38°}{2}} \\
&= 583.02 \text{ ft} \quad (580 \text{ ft})
\end{aligned}
$$

The answer is (C).

Why Other Options Are Wrong

(A) This incorrect answer would be the radius if the angles were not divided by two when calculating the radius.

(B) This answer incorrectly assumes that the long-chord formula would apply, using line AB as the chord.

(D) By inadvertently multiplying the tangent values in the denominator instead of adding them, the radius formula produces an overly large radius.

SOLUTION 3

Construct a triangle by connecting point O (the center) to point P, then drop an altitude from point P, intersecting base AO at point E.

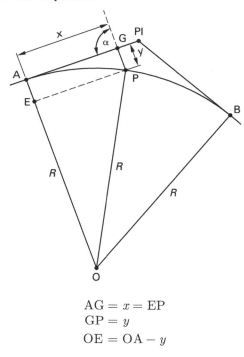

$$AG = x = EP$$
$$GP = y$$
$$OE = OA - y$$

From right triangle OEP,

$$OE = \sqrt{R^2 - x^2}$$

Find the curve radius.

$$
\begin{aligned}
D &= \frac{(180°)(100)}{\pi R} \\
R &= \frac{(180°)(100)}{\pi D} = \frac{(180°)(100 \text{ ft})}{\pi 2°} \\
&= 2864.79 \text{ ft} \\
OE &= \sqrt{(2864.79 \text{ ft})^2 - (200 \text{ ft})^2} \\
&= 2857.80 \text{ ft} \\
y &= OA - OE \\
&= 2864.79 \text{ ft} - 2857.80 \text{ ft} \\
&= 6.99 \text{ ft} \quad (7.0 \text{ ft})
\end{aligned}
$$

The answer is (C).

Why Other Options Are Wrong

(A) Determining the curve radius by incorrectly applying the conversion from degree of curve to radius will result in a radius that is much too large.

(B) This incorrect answer results from confusing the curve deflection to point P with the degree of curve. The deflection from AG to the chord AP is one half of the angle AOP, incorrectly assumed to be 2°.

(D) This incorrect answer is the result of using the long chord and mid-ordinate formulas to solve for y and incorrectly doubling the deflection angle.

SOLUTION 4

With the deflection and external distance known, solve for the first radius. Use the formula that relates the external distance to the curve radius and curve deflection.

$$E = R\left(\frac{1}{\cos\dfrac{I}{2}} - 1\right)$$

Rearrange.

$$R = \frac{E}{\dfrac{1}{\cos\dfrac{I}{2}} - 1} = \frac{35 \text{ ft}}{\dfrac{1}{\cos\dfrac{67°45'}{2}} - 1}$$
$$= 171.192 \text{ ft}$$

Find the degree of curve.

$$D = \frac{(180°)(100)}{\pi R} = \frac{(180°)(100 \text{ ft})}{\pi(171.192 \text{ ft})}$$
$$= 33.469°$$

Since the nearest whole degree is 33°, find the radius for a 33° curve.

$$R = \frac{5729.58}{D} = \frac{5729.578° \text{-ft}}{33°}$$
$$= 173.624 \text{ ft} \quad (170 \text{ ft})$$

The answer is (B).

Why Other Options Are Wrong

(A) Determining the secant from the inverse of the sine (instead of from the inverse of the cosine) will result in this incorrect answer.

(C) This incorrect answer is the radius before rounding the deflection to an even 33°.

(D) Incorrectly using the formula for the mid-ordinate (instead of the external distance) results in this answer.

SOLUTION 5

Solve for the length of the curve tangent.

$$T = R\tan\frac{I}{2} = (2500 \text{ ft})\left(\tan\frac{35°}{2}\right)$$
$$= 788.25 \text{ ft}$$

Find the station of the point of curvature (PC).

$$\begin{array}{r} \text{PI sta } 485 + 26.75 \\ - \quad 7 \quad\quad 88.25 \\ \hline \text{PC sta } 477 + 38.50 \end{array}$$

The length of the curve is added to the PC station to determine the station of the PT. Solve for length of curve.

$$L = RI\left(\frac{2\pi}{360°}\right) = (2500 \text{ ft})(35°)\left(\frac{2\pi}{360°}\right)$$
$$= 1527.16 \text{ ft}$$

Find the station of the PT.

$$\begin{array}{r} \text{PC sta } 477 + 38.50 \\ + \quad 15 \quad\quad 27.16 \\ \hline \text{PC sta } 492 + 65.66 \quad (\text{PC sta } 492 + 65) \end{array}$$

The answer is (C).

Why Other Options Are Wrong

(A) Using the full deflection angle instead of dividing by 2 in the tangent length equation will produce the wrong PC station.

(B) Using the formula for the chord length and adding that dimension to the PC station results in a station that is too low, because the stationing will not follow the centerline.

(D) Adding the tangent length to the PI results in a possible second station of the PT, but this is not correct practice.

SOLUTION 6

As recommended by GDHS, Eq. 3-7 is used for determining the maximum comfortable speed on horizontal curves.

$$f_{\max} = \frac{\text{v}^2}{15R} - \frac{e}{100\%}$$

Rearrange to find R.

$$R = \frac{\text{v}^2}{15\left(\dfrac{e}{100\%} + f_{\max}\right)}$$

The maximum friction factor, $f_{\max}$, is shown in Fig. 3-6 to be 0.20 for a design speed of 30 mph.

Solve for the radius.

$$R = \frac{\left(30\,\dfrac{\text{mi}}{\text{hr}}\right)^2}{\left(15\,\dfrac{\text{mi}^2}{\text{hr}^2\text{-ft}}\right)\left(\dfrac{-4\%}{100\%} + 0.20\right)}$$
$$= 375 \text{ ft}$$

Alternate Solution

From Table 3.13(b), for a design speed of 30 mph and $e = -4\%$, select 375 ft as the minimum radius.

The answer is (D).

Why Other Options Are Wrong

(A) This incorrect answer results from entering the cross slope directly in percent without converting to feet per foot, and then ignoring the negative sign in the answer.

(B) This incorrect answer results from referring to Table 3.13(a), using the column titled $\text{v} = 30$ km/h, and converting the answer from meters to feet.

(C) This incorrect answer results from ignoring the negative sign for the cross slope.

SOLUTION 7

Find the curve tangent.

$$T = R\tan\frac{I}{2} = (1400 \text{ ft})\left(\tan\frac{32°20'}{2}\right)$$
$$= 405.85 \text{ ft}$$

Determine the PT coordinates using the tangent bearing and distance from the point of intersection (PI) coordinates.

$$N_{\text{PT}} = N_{\text{PI}} + T\sin(90° - \text{bearing angle})$$
$$= 5280 \text{ ft} + (405.85 \text{ ft})\sin(90° - 55°40')$$
$$= 5508.90 \text{ ft} \quad (5510 \text{ ft})$$

The answer is (C).

Why Other Options Are Wrong

(A) This incorrect answer is the result of calculating the east coordinate instead of the north coordinate.

(B) This incorrect answer results from using the curve deflection angle to locate the PT coordinates from the PI instead of using the bearing of the ahead tangent.

(D) This incorrect answer results from neglecting to subtract the tangent bearing from 90° when determining the PT coordinate.

SOLUTION 8

Set up triangle AOC using the center of the curve as point O. Point A is the curve PC, and point C is the point on the curve that is radial from the pier corner, point F.

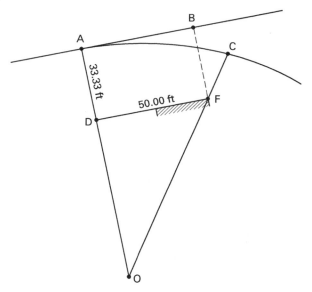

Find the distance OD.

$$\text{OD} = 200 \text{ ft} - 33.33 \text{ ft}$$
$$= 166.67 \text{ ft}$$

Find the distance OF using the Pythagorean theorem.

$$\text{OF} = \sqrt{(\text{DF})^2 + (\text{OD})^2} = \sqrt{(50.00 \text{ ft})^2 + (166.67 \text{ ft})^2}$$
$$= 174.01 \text{ ft}$$

Find the distance FC by subtracting OF (174.01 ft) from the curve radius.

$$\text{FC} = 200.00 \text{ ft} - 174.01 \text{ ft}$$
$$= 25.99 \text{ ft} \quad (26 \text{ ft})$$

The answer is (A).

Horizontal
Design

Why Other Options Are Wrong

(B) This incorrect answer results from using the formula for the offset from tangent to a point on a curve, then subtracting the result from the offset to the pier foundation corner.

(C) This incorrect answer results from considering triangle AOB. The hypotenuse, BO, is 206.16 ft. Subtracting the offset of 33.33 ft from the hypotenuse length yields 34.36 ft, which is the incorrect clearance distance to the closest corner of the pier foundation.

(D) This incorrect answer results from subtracting the squares of the side lengths in the Pythagorean theorem instead of adding the squares of the sides.

SOLUTION 9

Stopping distance around an obstruction is the distance along the centerline of the innermost lane of the curve, which is on a 250 ft radius. The sight line is along the chord of the curve. The clearance distance, $Z_{\text{clearance}}$, plus one-half of the inside lane width, is the horizontal sight offset (HSO). The HSO is the mid-ordinate, M, of the circular curve centered at the point of closest clearance to the edge of the road.

$$\text{HSO} = M = Z_{\text{clearance}} + \frac{\text{lane width}}{2}$$
$$= 14 \text{ ft} + \frac{12 \text{ ft}}{2}$$
$$= 20 \text{ ft}$$

Determine the curve length using the mid-ordinate and the radius. First, find the deflection, I, of the curve arc. The deflection is the angle subtended by the ends of the sight-line chord.

$$M = R\left(1 - \cos\frac{I}{2}\right)$$

Rearrange to solve for I.

$$I = 2\arccos\left(1 - \frac{M}{R}\right)$$
$$= 2\arccos\left(1 - \frac{20 \text{ ft}}{250 \text{ ft}}\right)$$
$$I = 46.148°$$

Determine the length of curve, L, using the radius and the degree of curve, D.

$$L = \frac{100I}{D} = \frac{\pi IR}{180°} = \frac{\pi(46.148°)(250 \text{ ft})}{180°}$$
$$= 201.4 \text{ ft}$$

Determine the chord length for the sight distance.

$$C = 2R\sin\frac{I}{2} = (2)(250 \text{ ft})(\sin 46.148°)$$
$$= 195.56 \text{ ft}$$

A sight distance of 195.96 ft is good for a 30 mph design speed.

The answer is (C).

Alternate Solution

Using AASHTO's *A Policy on Geometric Design of Highways and Streets* (GDHS) Fig. 3-26, from a point at 20 ft mid-ordinate on the bottom horizontal scale, intersect a line from a point at a 250 ft radius on the left vertical scale. The intersection lies on the v = 30 mph curve.

Why Other Options Are Wrong

(A) After determining $I/2$ from the mid-ordinate equation, this incorrect answer results from not doubling the answer for the length-of-curve equation.

(B) This answer results from incorrectly using the radius of the inside edge of the lane as the sight line.

(D) This answer results from incorrectly using the radius of the outside edge of the traveled lane as the sight line.

SOLUTION 10

Change the design speed from miles per hour to feet per second.

$$v_{\text{ft/sec}} = \frac{\left(70 \frac{\text{mi}}{\text{hr}}\right)\left(5280 \frac{\text{ft}}{\text{mi}}\right)}{3600 \frac{\text{sec}}{\text{hr}}}$$
$$= 102.7 \text{ ft/sec}$$

The roadway is already sloped down to the right at 0.01 ft/ft, and the superelevation will increase the slope of the roadway down to the right since this is a right-hand curve. Therefore, there will be no reverse cross slope to run out on the right side. Determine the difference between normal slope and full superelevation.

$$\text{required slope transition} = 0.08 \frac{\text{ft}}{\text{ft}} - 0.01 \frac{\text{ft}}{\text{ft}}$$
$$= 0.07 \text{ ft/ft}$$

The number of seconds required to transition from normal slope to full superelevation is determined by dividing that transition slope change by the allowable slope change rate.

$$
\begin{aligned}
\text{min transition time} &= \frac{\text{required slope transition}}{\text{slope change rate}} \\
&= \frac{0.07 \, \dfrac{\text{ft}}{\text{ft}}}{0.02 \, \dfrac{\text{ft}}{\text{ft-sec}}} \\
&= 3.5 \text{ sec}
\end{aligned}
$$

The length of superelevation must be at least the minimum transition time multiplied by the design speed.

$$
\begin{aligned}
L_{\min} &= (\text{min transition time})(\text{speed}) \\
&= (3.5 \text{ sec})\left(102.7 \, \frac{\text{ft}}{\text{sec}}\right) \\
&= 359.3 \text{ ft} \quad (360 \text{ ft})
\end{aligned}
$$

The answer is (B).

Why Other Options Are Wrong

(A) This incorrect answer is the distance traveled in one second, which is an insufficient length to obtain the full superelevation.

(C) In this incorrect answer, transitioning the full 0.08 ft/ft and ignoring the pavement normal cross slope of 0.01 ft/ft results in too long of a transition.

(D) In this incorrect answer, including the reverse-slope runout in the transition length results in too long of a distance.

SOLUTION 11

The highway centerline radius, R_h, is determined as follows.

$$
R_h = R_{\mathrm{rr}} - 125 \text{ ft}
$$

Convert the railroad curve to the radius definition using the chord definition formula.

$$
R_{\mathrm{rr}} = \frac{50}{\sin \dfrac{D}{2}} = \frac{50°\text{-ft}}{\sin \dfrac{6°}{2}}
$$
$$
= 955.37 \text{ ft}
$$

Determine the highway centerline radius.

$$
\begin{aligned}
R_h &= R_{\mathrm{rr}} - 125.00 \text{ ft} \\
&= 955.37 \text{ ft} - 125 \text{ ft} \\
&= 830.37 \text{ ft}
\end{aligned}
$$

Determine the highway degree of curve using the arc definition formula.

$$
\begin{aligned}
D &= \frac{(180°)(100)}{\pi R} = \frac{(180°)(100 \text{ ft})}{\pi (830.37 \text{ ft})} \\
&= 6.9000° \quad (6°54'00'')
\end{aligned}
$$

The answer is (B).

Why Other Options Are Wrong

(A) This incorrect answer results from placing the railroad curve to the inside of the highway (i.e., adding 125 ft to the railroad radius).

(C) This incorrect answer results from using the arc definition for the railroad curve.

(D) This incorrect answer results from not dividing D by 2 in the chord definition equation.

SOLUTION 12

Determine the length of the back tangent (from the point of curvature (PC) to the point of intersection (PI)) by the difference of coordinates.

$$
\begin{aligned}
T &= \sqrt{(N_{\mathrm{PI}} - N_{\mathrm{PC}})^2 + (E_{\mathrm{PI}} - E_{\mathrm{PC}})^2} \\
&= \sqrt{(5280 \text{ ft} - 5280 \text{ ft})^2 + (3546 \text{ ft} - 2640 \text{ ft})^2} \\
&= 906 \text{ ft}
\end{aligned}
$$

Find the curve deflection.

$$
T = R \tan \frac{I}{2}
$$

Rearrange to find I.

$$
\begin{aligned}
I &= 2 \arctan \frac{T}{R} \\
&= 2 \arctan \frac{906 \text{ ft}}{2300 \text{ ft}} \\
&= 43°
\end{aligned}
$$

Find the bearing of the ahead tangent (PI to point of tangent (PT)). The back tangent (PC to PI) lies in a due-east position. Bearings are taken from due south or due north.

$$
90° - 43° = 47°
$$

The ahead tangent will be in the southeasterly quadrant. The bearing is S 47° E.

The answer is (C).

Why Other Options Are Wrong

(A) This incorrect answer is the azimuth of the ahead tangent.

(B) This incorrect answer is the result of using the curve deflection angle as the bearing angle.

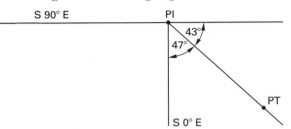

(D) This incorrect answer results from neglecting to double the angle obtained from the arctangent T/R.

SOLUTION 13

The roadway is already sloped down to the right at 0.01 ft/ft, and the superelevation will increase the slope of the roadway down to the right, since this is a right-hand curve. The difference between normal slope and full superelevation is found by subtracting the normal slope from the full slope.

$$e_{\text{full}} - e_{\text{normal}} = 0.08 \ \frac{\text{ft}}{\text{ft}} - 0.01 \ \frac{\text{ft}}{\text{ft}}$$
$$= 0.07 \ \text{ft/ft}$$

Determine the right-edge profile elevation change.

$$\Delta e = (\text{pavement width})(\text{slope change})$$
$$= (24 \ \text{ft})\left(0.07 \ \frac{\text{ft}}{\text{ft}}\right)$$
$$= 1.68 \ \text{ft}$$

Use the edge transition rate of 1:200 to determine the transition length.

$$L = \Delta e (\text{change rate})$$
$$= (1.68 \ \text{ft})\left(200 \ \frac{\text{ft}}{\text{ft}}\right)$$
$$= 336 \ \text{ft} \quad (340 \ \text{ft})$$

The answer is (B).

Why Other Options Are Wrong

(A) This incorrect answer is the transition rate, not the transition distance.

(C) In this incorrect answer, transitioning the full 0.08 ft/ft would result in too long of a transition.

(D) This incorrect answer results from improper interpretation of the transition-rate definition.

$$\frac{200 \ \text{ft}}{0.07 \ \dfrac{\text{ft}}{\text{ft}}} = 2857 \ \text{ft}$$

SOLUTION 14

Superelevation, e, of a railroad track is the amount of rise (elevation) given to the outer rail of a curve at the gage line over the inner rail. The transition length, s, is determined by dividing the superelevation of the curve, e, by the allowable rate of change in inches per second, Δe, times the velocity.

$$s = \left(\frac{e}{\Delta e}\right)\text{v}$$

The rate-of-change angle is converted to the rate of change in elevation using the tangent of the rate-of-change angle. The tangent of the rate-of-change cross-slope angle is the elevation change of the high rail, Δe, divided by the track gage, 56.5 in.

$$\tan 1.15° = 0.02$$

Determine the rate of change, Δe, in inches per second.

$$\Delta e = \left(0.02 \ \frac{1}{\text{sec}}\right)(56.5 \ \text{in})$$
$$= 1.13 \ \text{in/sec}$$

Determine the number of seconds, t, to attain full superelevation.

$$t = \frac{e}{\Delta e} = \frac{5 \ \text{in}}{1.13 \ \dfrac{\text{in}}{\text{sec}}}$$
$$= 4.41 \ \text{sec}$$

Determine the distance traveled in 4.41 sec.

$$s = \mathrm{v}t$$
$$= \frac{\left(150 \ \frac{\text{mi}}{\text{hr}}\right)\left(5280 \ \frac{\text{ft}}{\text{mi}}\right)(4.41 \ \text{sec})}{3600 \ \frac{\text{sec}}{\text{hr}}}$$
$$= 970 \ \text{ft}$$

The answer is (C).

Why Other Options Are Wrong

(A) This incorrect answer results from dividing the full superelevation by the tangent of the rate-of-change angle.

(B) This incorrect answer results from a faulty speed conversion.

(D) This incorrect answer results from dividing the speed by the tangent of the rate-of-change angle.

SOLUTION 15

Hickerson's *Route Location and Design* uses the following formula for obtaining the external distance of the curve.

$$E = \left(R + p\right)\sec\frac{I}{2} - R$$

Find the secant.

$$\sec\frac{I}{2} = \frac{1}{\cos\dfrac{I}{2}} = \frac{1}{\cos\dfrac{24°}{2}}$$
$$= \frac{1}{\cos 12°}$$

Find y.

$$y = \frac{L_s^2}{6R} = \frac{(250 \ \text{ft})^2}{(6)(850 \ \text{ft})}$$
$$= 12.25 \ \text{ft}$$

Find the spiral angle.

$$I_s = \left(\frac{L_s}{200}\right)\left(\frac{(180°)(100)}{\pi R}\right)$$
$$= \left(\frac{250 \ \text{ft}}{200 \ \text{ft}}\right)\left(\frac{(180°)(100 \ \text{ft})}{\pi(850 \ \text{ft})}\right)$$
$$= 8.42585°$$

Solve for p.

$$p = y - R(1 - \cos I_s)$$
$$= 12.25 \ \text{ft} - (850 \ \text{ft})(1 - \cos 8.42585°)$$
$$= 3.08 \ \text{ft}$$

$$E = (850 \ \text{ft} + 3.08 \ \text{ft})\left(\frac{1}{\cos 12°}\right) - 850 \ \text{ft}$$
$$= 22.14 \ \text{ft} \quad (22 \ \text{ft})$$

The answer is (B).

Why Other Options Are Wrong

(A) In this incorrect answer, not including the p distance in the external distance equation yields an answer that is too small.

(B) In this incorrect answer, using the value for y instead of the value for p in the external distance equation yields an answer that is too large.

(D) In this incorrect answer, not dividing the deflection by two in the external distance equation yields an answer that is too large.

SOLUTION 16

Determine the curve radius.

$$R = \frac{(180°)(100)}{\pi D} = \frac{(180°)(100 \ \text{ft})}{(\pi)(5.73°)}$$
$$= 1000 \ \text{ft}$$

Determine the spiral length using the given formula.

$$L_s = \left(1.6 \ \frac{\text{ft}^2\text{-hr}^3}{\text{mi}^3}\right)\left(\frac{\text{v}_{\text{mph}}^3}{R_{\text{ft}}}\right)$$
$$= \left(1.6 \ \frac{\text{ft}^2\text{-hr}^3}{\text{mi}^3}\right)\left(\frac{\left(60 \ \frac{\text{mi}}{\text{hr}}\right)^3}{1000 \ \text{ft}}\right)$$
$$= 346 \ \text{ft} \quad (350 \ \text{ft})$$

Determine the change in cross slope.

$$\text{change in cross slope} = \text{full slope} - \text{normal slope}$$
$$= 0.08 \ \frac{\text{ft}}{\text{ft}} - 0.015 \ \frac{\text{ft}}{\text{ft}}$$
$$= 0.065 \ \text{ft/ft}$$

Determine the change in edge of pavement elevation for a three-lane half width of roadway.

$$\Delta e = ew$$
$$= \left(0.065 \ \frac{\text{ft}}{\text{ft}}\right)\left(12 \ \frac{\text{ft}}{\text{lane}}\right)(3 \ \text{lanes})$$
$$= 2.34 \ \text{ft}$$

Use the edge transition rate of 1:200 to determine the transition length.

$$L_{tr} = (\Delta e) G_r$$
$$= \left(2.34 \ \text{ft}\right)\left(200 \ \frac{\text{ft}}{\text{ft}}\right)$$
$$= 468 \ \text{ft} \quad (470 \ \text{ft})$$

The slope ratio transition length, 468 ft, is greater than the spiral formula length, 346 ft. Using the longer of the two criteria, set the spiral length to 470 ft.

The answer is (B).

Why Other Options Are Wrong

(A) This incorrect answer is the spiral length using the spiral formula, which is shorter than the edge transition criterion.

(C) This incorrect answer is the result of using the full slope transition without considering that the pavement is already sloped 0.015 ft/ft at the beginning of the transition.

(D) This incorrect answer results from using the entire six-lane width to determine the edge transition elevation.

SOLUTION 17

The design speed determines the distance necessary to change from one curve radius to the next, which gives the driver time to adjust to the new radius. When calculating spiral length, AASHTO's *A Policy on Geometric Design of Highways and Streets* (GDHS), recommends keeping the change in lateral acceleration in the range of 1.0–3.0 ft/sec², with most cases being satisfied using an average of 2.0 ft/sec². *GDHS* Eq. 3-25 can be used for transition between curves of any radius, in addition to the traditional tangent to curve condition. The length of spiral required is the length necessary to transition from the equivalent radius of a 3° curve to a 2° curve. The radius of a curve is

$$R = \frac{(180°)(100 \ \text{ft})}{\pi D}$$

Solve for each curve.

$$R_1 = \frac{(180°)(100 \ \text{ft})}{\pi(3°)}$$
$$= 1909.86 \ \text{ft}$$
$$R_2 = \frac{(180°)(100 \ \text{ft})}{\pi(2°)}$$
$$= 2864.79 \ \text{ft}$$

Determine the length of the transition by finding the distance along the spiral in which the radius changes from R_1 to R_2.

$$L_s = L_1 - L_2 = \frac{1.6v^3}{R_1} - \frac{1.6v^3}{R_2}$$
$$= \left(\frac{\left(1.6 \ \frac{\text{ft}^2\text{-hr}^3}{\text{mi}^3}\right)\left(70 \ \frac{\text{mi}}{\text{hr}}\right)^3}{1909.86 \ \text{ft}}\right)$$
$$- \left(\frac{\left(1.6 \ \frac{\text{ft}^2\text{-hr}^3}{\text{mi}^3}\right)\left(70 \ \frac{\text{mi}}{\text{hr}}\right)^3}{2864.79 \ \text{ft}}\right)$$
$$= 95.78 \quad (100 \ \text{ft})$$

The answer is (A).

Why Other Options Are Wrong

(B) This incorrect answer is the length of spiral needed at the tangent end of a 2° curve.

(C) This incorrect answer is the length of spiral needed at the tangent end of a 3° curve.

(D) This incorrect answer is the length of spiral needed if a full spiral were introduced between the curves that transitioned to tangent and then transitioned into the second curve. This method is not a true compound spiral curve but rather two fully spiraled curves back-to-back.

SOLUTION 18

The formula given for the spiral length in the problem is from AASHTO's *A Policy on Geometric Design of Highways and Streets* (GDHS). It is based on speed in miles per hour and radius in feet, when the change in lateral acceleration is 2.0 ft/sec². The curve radius calculation using degree of curve can be incorporated into the base equation by substituting for R.

$$R = \frac{(360°)(100)}{2\pi D}$$

$$L_s = 1.6v^3\left(\frac{2\pi D}{(360°)(100)}\right)$$

Determine the change in degree of curve.

$$D = D_1 - D_2 = 2° - (-3°)$$
$$= 5°$$

Determine the length of the transition spiral required.

$$L_s = \left(1.6\ \frac{\text{ft}^2\text{-hr}^3}{\text{mi}^3}\right)\left(60\ \frac{\text{mi}}{\text{hr}}\right)^3\left(\frac{2\pi(5°)}{(360°)(100\ \text{ft})}\right)$$
$$= 301.6\ \text{ft} \quad (300\ \text{ft})$$

The answer is (D).

Why Other Options Are Wrong

(A) This incorrect answer is the required length of spiral for a 1° change in degree of curve. This solution would be true for a compound curve but is incorrect for a reverse curve.

(B) This incorrect answer is the required length of spiral to bring a 2° curve back to zero deflection. More spiral length is required to develop the 3° curve in the opposite direction.

(C) This incorrect answer is the required length of spiral to develop a 3° curve from zero deflection (tangent). More spiral length is required to bring the 2° curve deflection to zero.

SOLUTION 19

The spiral offset is defined by

$$p = y_s - R\,\text{vers}\,I_s$$

Determine the curve radius.

$$R = \frac{(180°)(100)}{\pi D} = \frac{(180°)(100\ \text{ft})}{(\pi)(5°)}$$
$$= 1145.92\ \text{ft}$$

Determine the spiral angle.

$$I_s = \left(\frac{L_s}{200}\right)D = \left(\frac{200\ \text{ft}}{200\ \text{ft}}\right)(5°)$$
$$= 5°$$

Determine the tangent offset.

$$y_s = \frac{L_s^2}{6R} = \frac{(200\ \text{ft})^2}{(6)(1145.92\ \text{ft})}$$
$$= 5.82\ \text{ft}$$

Determine the spiral offset.

$$p = 5.82\ \text{ft} - (1145.92\ \text{ft})\text{vers}\ 5°$$
$$= 1.46\ \text{ft}$$

The answer is (B).

Why Other Options Are Wrong

(A) This incorrect value results from using the sine of the spiral deflection instead of the versine in the spiral offset equation.

(C) This incorrect answer is the external distance of the full curve, which indicates how much the point of intersection (PI) is shifted radially from the unspiraled curve.

(D) This incorrect answer is the tangent offset, y_s, of the full spiral at the spiral-to-curve point (SC).

3 Vertical Design

PROBLEM 1

A -4% grade meets a $+5\%$ grade at sta 34+00.

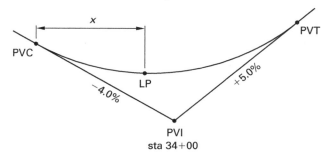

Using a 600 ft vertical curve, what is most nearly the distance from the point of vertical curve (PVC) to the low point (LP)?

(A)   0.44 ft ahead from the PVC

(B)   2.7 ft ahead from the PVC

(C)   44 ft ahead from the PVC

(D)   270 ft ahead from the PVC

Hint: The low point is where the grade is equal to zero.

PROBLEM 2

A 500 ft long sag vertical curve passes under a bridge at sta 82+45. The point of vertical curve (PVC) is at sta 81+00. A -3.6% grade meets a $+4.4\%$ grade at the point of vertical intersection (PVI), which is at elevation 425.38 ft.

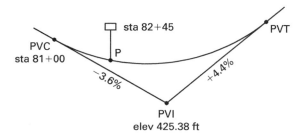

What is most nearly the elevation of the point under the bridge?

(A)   413 ft

(B)   430 ft

(C)   431 ft

(D)   441 ft

Hint: The elevation of any point on a vertical curve can be found as long as one other elevation point on the curve is known.

PROBLEM 3

A vertical curve is to be placed through sta 81+50 at an elevation of 638.42 ft. The grade into the curve is -3.6%, and the grade out of the curve is $+3.8\%$. The point of vertical curve (PVC) is at sta 76+00 and elevation 646.12 ft.

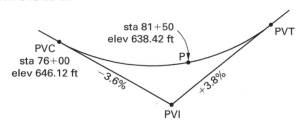

What is most nearly the required length of curve?

(A)   25.0 ft

(B)   168 ft

(C)   407 ft

(D)   925 ft

Hint: Curve elevations on parabolic vertical curves are found by comparison with other known elevation points. The relationship can be worked inversely to determine curve length if two points are already known.

Vertical Design

PROBLEM 4

A specification for a high-speed railway states that vertical acceleration in a parabolic sag vertical curve is to be limited to 0.03 g. Using a design speed of 150 mph, a −0.75% grade meets a +0.45% grade. What is most nearly the minimum length of vertical curve required?

(A) 150 ft

(B) 280 ft

(C) 600 ft

(D) 1980 ft

Hint: The change in vertical velocity is related to time and travel speed when calculating the distance necessary to change vertical direction.

PROBLEM 5

An interchange entrance ramp on a rural freeway at the far end of a 1500 ft crest vertical curve has been the scene of several rear-end crashes. The terrain is mountainous, and the road surface is often wet in foggy conditions. At this location, the freeway has a 3.0% upgrade meeting a 2.5% downgrade. What should be the posted maximum speed for the freeway?

(A) 50 mph

(B) 60 mph

(C) 65 mph

(D) 70 mph

Hint: The sight distance on a crest vertical curve follows American Association of State Highway and Transportation Officials (AASHTO) guidelines.

PROBLEM 6

A 3% upgrade meets a 4% downgrade on a road with a design speed of 50 mph. What is most nearly the minimum length of vertical curve required based on American Association of State Highway and Transportation Officials (AASHTO) recommendations for stopping sight distance?

(A) 425 ft

(B) 540 ft

(C) 590 ft

(D) 670 ft

Hint: Two situations can occur—one in which the stopping sight distance is shorter than the length of vertical curve, and the other in which the stopping sight distance is longer than the vertical curve.

PROBLEM 7

A sag vertical curve with no passing permitted is being rebuilt on a roadway in an unlit suburban area. The grade entering the curve is −6% leading into a +3.5% grade. Using the sag vertical curve formula from *A Policy on Geometric Design of Highways and Streets* (GDHS) and a speed limit of 35 mph, what is most nearly the required length of curve?

(A) 130 ft

(B) 250 ft

(C) 500 ft

(D) 1000 ft

Hint: The GDHS formula is based on headlight sight distance. Only stopping sight distance should be considered, since no passing is permitted.

SOLUTION 1

The rate of change of grade is uniform along the curve. The point where the grade is zero will be proportional to the distance along the curve related to the grade change for the total curve length.

$$x = \frac{G_1 L}{G_1 - G_2} = \frac{(-4\%)(6 \text{ sta})}{-4\% - 5\%}$$
$$= 2.667 \text{ sta} \quad (270 \text{ ft})$$

The low point is approximately 270 ft ahead from the PVC.

The answer is (D).

Why Other Options Are Wrong

(A) In this incorrect answer, the proportional distance along the curve is incomplete because it was not multiplied by the curve length.

(B) This incorrect answer is the station distance, misconstrued as feet.

(C) In this incorrect answer, the proportional distance along the curve was not multiplied by the curve length and was converted to feet.

SOLUTION 2

The length of the curve is 500 ft, which is 5 stations. Find the elevation of the PVC.

$$E_{\text{PVC}} = E_{\text{PVI}} + G_1 \left(\frac{L}{2} \right)$$
$$= 425.38 \text{ ft} + \left(3.6 \frac{\text{ft}}{\text{sta}} \right) \left(\frac{5 \text{ sta}}{2} \right)$$
$$= 434.38 \text{ ft}$$

Find the rate of change of grade per station.

$$R = \frac{G_2 - G_1}{L} = \frac{4.4\% - (-3.6\%)}{5 \text{ sta}}$$
$$= 1.60\%/\text{sta}$$

Use that rate to determine the elevation of the point under the overpass. The elevation of the PVC is used as the known point in the equation.

$$E_P = E_{\text{PVC}} + G_1 x + \left(\frac{R}{2} \right) x^2$$
$$= 434.38 \text{ ft}$$
$$+ \left(\begin{pmatrix} -0.036 \frac{\text{ft}}{\text{ft}} \end{pmatrix} (82.45 \text{ sta} - 81.00 \text{ sta}) \\ + \left(\frac{0.016 \frac{\text{ft}}{\text{ft}}}{2 \text{ sta}} \right) (82.45 \text{ sta} - 81.00 \text{ sta})^2 \right)$$
$$\times \left(100 \frac{\text{ft}}{\text{sta}} \right)$$
$$= 430.84 \text{ ft} \quad (431 \text{ ft})$$

The answer is (C).

Why Other Options Are Wrong

(A) In this incorrect answer, the elevation of the PVC was incorrectly determined.

(B) In this incorrect answer, the distance term was not squared.

(D) In this incorrect answer, in the second term of the elevation equation, the negative sign in front of the 3.6% grade was missed.

SOLUTION 3

Determine the distance between the two known points.

$$x = 8150 \text{ ft} - 7600 \text{ ft} = 550 \text{ ft}$$

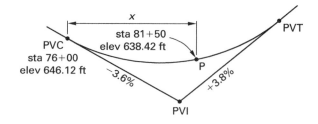

The rate of change can be related to the unknown length of the vertical curve.

$$R = \frac{G_2 - G_1}{L} = \frac{0.038 \frac{\text{ft}}{\text{ft}} - \left(-0.036 \frac{\text{ft}}{\text{ft}}\right)}{L}$$

$$= \frac{0.074 \frac{\text{ft}}{\text{ft}}}{L}$$

$$\frac{R}{2} = \frac{0.037 \frac{\text{ft}}{\text{ft}}}{L}$$

Insert values into the complete elevation formula.

$$E_P = E_{PVC} + G_1 x + \left(\frac{R}{2}\right) x^2$$

$$638.42 \text{ ft} = 646.12 \text{ ft} + \left(-0.036 \frac{\text{ft}}{\text{ft}}\right)(550 \text{ ft})$$

$$+ \left(\frac{0.037 \frac{\text{ft}}{\text{ft}}}{L}\right)(550 \text{ ft})^2$$

Rearrange.

$$L = \frac{\left(0.037 \frac{\text{ft}}{\text{ft}}\right)(550 \text{ ft})^2}{638.42 \text{ ft} - 642.12 \text{ ft} + \left(0.036 \frac{\text{ft}}{\text{ft}}\right)(550 \text{ ft})}$$

$$= 925 \text{ ft}$$

The answer is (D).

Why Other Options Are Wrong

(A) Incorrectly determining the difference in absolute grades instead of the total change in grade results in a very short grade length.

(B) Neglecting to square the distance in the third term and assuming the answer is in stations because it is so small results in a vertical curve that is much too short.

(C) Omitting the minus sign for G_1 in the second term of the elevation equation and then ignoring the minus sign of the answer results in a curve that is too short.

SOLUTION 4

Convert travel speed to feet per second.

$$v_{horiz} = \frac{\left(150 \frac{\text{mi}}{\text{hr}}\right)\left(5280 \frac{\text{ft}}{\text{mi}}\right)}{3600 \frac{\text{sec}}{\text{hr}}}$$

$$= 220 \text{ ft/sec}$$

Using 1 g = 32.2 ft/sec^2, calculate the vertical acceleration limit.

$$a_{vert} = (0.03)\left(32.2 \frac{\text{ft}}{\text{sec}^2}\right)$$

$$= 0.966 \text{ ft/sec}^2$$

When the train is traveling on a downgrade, it has a negative vertical velocity. On an upgrade, the train has a positive vertical velocity.

Determine the vertical velocity on the downgrade.

$$v_{vert} = G_1 v_{horiz} = \left(\frac{-0.75\%}{100\%}\right)\left(220 \frac{\text{ft}}{\text{sec}}\right)$$

$$= -1.650 \text{ ft/sec}$$

Determine the vertical velocity on the upgrade.

$$v_{vert} = G_2 v_{horiz} = \left(\frac{0.45\%}{100\%}\right)\left(220 \frac{\text{ft}}{\text{sec}}\right)$$

$$= 0.990 \text{ ft/sec}$$

Determine the total change in vertical velocity.

$$\Delta v = |v_{down} - v_{up}|$$

$$= \left|-1.650 \frac{\text{ft}}{\text{sec}} - 0.990 \frac{\text{ft}}{\text{sec}}\right|$$

$$= 2.64 \text{ ft/sec}$$

Determine the number of seconds required to change direction.

$$t = \frac{\Delta v}{a_{vert}} = \frac{2.64 \frac{\text{ft}}{\text{sec}}}{0.966 \frac{\text{ft}}{\text{sec}^2}}$$

$$= 2.73 \text{ sec}$$

The distance traveled in 2.73 sec is the minimum length of vertical curve required.

$$L_{\min} = t\text{v}_{\text{horiz}} = (2.73 \text{ sec})\left(220 \ \frac{\text{ft}}{\text{sec}}\right)$$
$$= 601 \text{ ft} \quad (600 \text{ ft})$$

The answer is (C).

Why Other Options Are Wrong

(A) Using the difference in absolute vertical velocity values results in too short a curve.

(B) Neglecting to convert miles per hour into feet per second results in too short a curve.

(D) Inserting the metric value for gravitational acceleration and not converting to feet per second squared to determine the vertical acceleration limit results in too long a curve.

SOLUTION 5

Check for the approximate required sight distance from AASHTO's *A Policy on Geometric Design of Highways and Streets* (GDHS) Table 3-1. The recommended stopping distance for 70 mph is 730 ft. Therefore, the curve length of 1500 ft is probably greater than the required stopping sight distance.

Check the available sight distance. Use the GDHS formula for crest vertical curves for stopping sight distances less than the curve length.

$$L = \frac{AS^2}{2158} = \frac{(G_2 - G_1)S^2}{2158}$$

Solve for the sight distance available.

$$S = \sqrt{\frac{2158L}{A}} = \sqrt{\frac{(2158 \text{ ft})(1500 \text{ ft})}{3.0\% - (-2.5\%)}}$$
$$= 767 \text{ ft}$$

From GDHS Table 3-3, 767 ft is in the lower range of 50 mph sight distance conditions. However, designing for minimum stopping sight distance on freeways is not entirely safe. A car braking rapidly or coming to a full stop can cause multiple rear-end crashes in heavier traffic found on freeways, especially when rain or fog are present. Therefore, avoidance maneuvers A and B do not apply, and one of the remaining conditions, C, D, or E, applies. Avoidance maneuvers C, D, and E allow for a greater perception-reaction time and allow time for the driver to change vehicle path or speed when a full

stop is undesirable. Avoidance maneuver C applies to rural roads; therefore, the speed should be posted at 50 mph.

The answer is (A).

Why Other Options Are Wrong

(B) Posting for 60 mph places the 767 ft stopping sight distance within the range of 610 ft to 1150 ft for avoidance manuevers A and B, which does not leave the additional distance required for lane change decisions and adverse weather conditions.

(C) Misinterpreting GDHS Table 3-2 by using the former criterion of a 0.6 ft object height yields a longer stopping sight distance required than does using the newer criteria. Applying the value to GDHS Table 3-2 yields a speed limit of 60 mph.

(D) Incorrectly using the stopping distance values from GDHS Table 3-1 and Table 3-2 for a design speed of 70 mph ignores the additional sight distance needed for avoidance maneuvers. Such measures may be required on the downgrade at the end of the vertical curve where the ramp merges into the mainline traffic, especially when experience has shown that a higher rate of crashes occur at this location.

SOLUTION 6

Assume $S < L$. From AASHTO's *A Policy on Geometric Design of Highways and Streets* (GDHS) Table 3-34, the stopping sight distance for a 50 mph design speed is 425 ft. Equation 3-43 is used with the eye height as 3.5 ft and the object height as 2.0 ft.

$$L = \frac{AS^2}{2158} = \frac{\left(3\% - (-4\%)\right)(425 \text{ ft})^2}{2158 \text{ ft}}$$
$$= 586 \text{ ft} \quad (590 \text{ ft})$$

Assume $S > L$, and solve for L.

$$L = 2S - \frac{2158}{A} = (2)(425 \text{ ft}) - \frac{2158 \text{ ft}}{\left(3\% - (-4\%)\right)}$$
$$= 542 \text{ ft} \quad (540 \text{ ft})$$

Since 542 ft is greater than the stopping sight distance required, the first assumption—that the stopping sight distance is less than the curve length—is valid. The length can be checked using the minimum design K value from GDHS Table 3-36.

$$L = KA = \left(84 \ \frac{\text{ft}}{\%}\right)\left(3\% - (-4\%)\right)$$
$$= 588 \text{ ft} \quad (590 \text{ ft})$$

The minimum required vertical curve length is 588 ft (590 ft).

The answer is (C).

Why Other Options Are Wrong

(A) This incorrect answer is the sight distance. It is wrong to assume that the curve length can be the same as the sight distance. A curve of this length would have a stopping sight distance of 360 ft, which is for a speed of 45 mph.

(B) Even though 540 ft would allow a shorter curve than the correct answer, the assumption that the sight distance is longer than the curve length is incorrect, making this curve length inadequate.

(D) Using an object height of 0.5 ft (from outdated GDHS criteria) yields a curve length that is greater than the minimum, as in this incorrect answer.

SOLUTION 7

Determine the stopping sight distance required for a design speed of 35 mph using GDHS Table 3-1. The recommended design distance is 250 ft.

Assume the curve length will be longer than the required sight distance.

$$
\begin{aligned}
L &= \frac{AS^2}{400 + 3.5S} \\
&= \frac{(G_2 - G_1)(250 \text{ ft})^2}{400 \text{ ft} + (3.5)(250 \text{ ft})} \\
&= \frac{\big(3.5\,\% - (-6\%)\big)(250 \text{ ft})^2}{400 \text{ ft} + (3.5)(250 \text{ ft})} \\
&= 466 \text{ ft} \quad (500 \text{ ft})
\end{aligned}
$$

Verify this result by using the formula for a curve length that is shorter than the required sight distance.

$$
\begin{aligned}
L &= 2S - \frac{400 + 3.5S}{A} \\
&= 2S - \frac{400 \text{ ft} + (3.5)(250 \text{ ft})}{G_2 - G_1} \\
&= (2)(250 \text{ ft}) - \frac{400 \text{ ft} + (3.5)(250 \text{ ft})}{3.5\% - (-6.0\%)} \\
&= 366 \text{ ft} \quad (370 \text{ ft})
\end{aligned}
$$

Both answers are greater than the required 250 ft. Therefore, the required length is 466 ft (500 ft).

The answer is (C).

Why Other Options Are Wrong

(A) Determining A to be the arithmetic instead of algebraic difference in grades results in too short a curve length using the $S < L$ formula and this incorrect answer.

(B) Using the stopping sight distance of 250 ft for the length of vertical curve results in too short a vertical curve and this incorrect answer.

(D) Determining A to be the arithmetic instead of algebraic difference in grades results in much too long a curve length using the $S > L$ formula and this incorrect answer.

Intersection Geometry

PROBLEM 1

Construction and operational characteristics of a full cloverleaf interchange between two freeways can be described by which of the following statements?

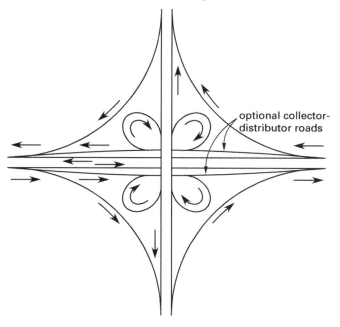

optional collector-distributor roads

(A) Only one bridge is necessary, making the interchange less costly to build.

(B) The layout makes the most efficient use of real estate and is easily understood by drivers.

(C) All movements can be made as a direct connection at nearly the same speed as mainline travel.

(D) All ramp entrances and exits are on the right side of the roadway, but may be too closely spaced to handle higher traffic volumes in the weaving zones.

Hint: A full cloverleaf interchange has four tight circular ramps tucked between broad outer ramps and the mainline.

PROBLEM 2

Beech St. is a 40 ft wide street through a residential neighborhood. Parking is permitted on both sides. Columbus Blvd. is a commercial multilane artery that runs parallel to Beech St. Congestion on Columbus Blvd. causes traffic to divert to Beech St. as a bypass. Cross streets, which are 38 ft wide with parking on both sides, occur frequently between Beech St. and Columbus Blvd. The location is subject to frequent snow and ice conditions.

Roundabouts are to be installed along Beech St. to discourage bypass traffic, especially large multi-trailer trucks. The design must accommodate school buses, municipal waste trucks, fire trucks, snow plows, and occasional WB-50 semi-trailer moving vans. The design should also consider that large semi-trailer rigs will occasionally stray onto Beech St. Assuming sufficient right-of-way is available, most nearly, what should be the minimum inscribed radius of the outside curb and central island treatment for the roundabouts?

(A) 30 ft; 10 ft diameter paved center island surrounded by a mountable curb

(B) 35 ft; landscaped center island surrounded by a barrier curb

(C) 45 ft; paved center island surrounded by a mountable curb

(D) 45 ft; no center island

Hint: Roundabouts are introduced to reduce left-turn conflicts and to moderate speed through intersections, replacing other traffic control measures.

PROBLEM 3

An office park will place pedestrian crosswalks in advance of intersection stop bars, as shown.

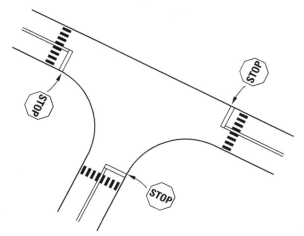

Which of the following statements are true for the design shown?

I.   The crosswalk lengths are shorter than those of crosswalks placed at curb returns, reducing the time of exposure to moving traffic.

II. Placing the crosswalks at the locations shown is more dangerous for pedestrians than if they were placed at the intersections.

III. Placing the crosswalks at the locations shown reduces pedestrian interference with turning traffic.

IV. The crosswalk locations are classified as midblock according to *Manual on Uniform Traffic Control Devices* (MUTCD) standards.

V.   The placement of the crosswalks in advance of intersection stop bars makes it unclear where drivers should stop for pedestrians.

VI. The placement of the crosswalks may encourage pedestrians to walk behind a vehicle stopped at the stop bar, thereby not being visible to drivers approaching from other directions.

(A) I, II, and III only

(B) II, IV, and V only

(C) I, II, IV, and VI only

(D) I, II, V, and VI only

Hint: Crosswalks at intersections are to be placed within the intersection according to design standards in MUTCD.

SOLUTION 1

The layout shows all mainline exits and entrances on the right side of the roadway.

The closeness of the inner entrance and exit ramps can make it more difficult to navigate through the weaving zone in heavy traffic. Accelerating entrance traffic will conflict with decelerating exit traffic.

The answer is (D).

Why Other Options Are Wrong

(A) The one bridge needed is very wide and very long but not necessarily less costly than several smaller bridges.

(B) Drivers may well understand the layout, but this interchange is more difficult to sign than are direct connection ramps. The large inner ramp loops consume much real estate that is landlocked by the ramps; therefore, this interchange does not make efficient use of the land.

(C) The only movements that are direct connections are the outer movements. Loop ramps are not considered direct connections. The smaller radius of the inner loop ramps requires considerable reduction in speed from the mainline.

SOLUTION 2

A mini-roundabout, as shown, would be best suited to the conditions described.

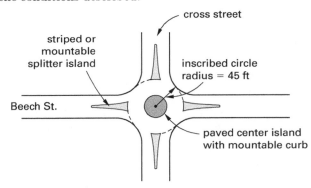

Mini-roundabouts have a maximum of one entering lane per approach, and have typical inscribed circle radii of 45 ft to 90 ft (90 ft diameter to 180 ft diameter). School buses, single-unit municipal service vehicles, and fire trucks can turn within an outside front corner radius of less than 45 ft under normal road conditions. Semi-trailer moving vans may not be able to turn without mounting the center island with the rear wheels, which is acceptable with a mountable curb island design. Additionally, the extra space provided by a 45 ft inscribed radius will allow for snow storage and small skidding actions of turning vehicles.

The answer is (C).

Why Other Options Are Wrong

(A) This incorrect answer depicts a 30 ft inscribed radius with a 10 ft paved center island. However, this does not provide sufficient room for maneuvering school buses and semi-trailer moving vans without the trailer wheels mounting the barrier curb. Also, stray large semi-trailer rigs cannot traverse the roundabout without potential damage to the curb.

(B) This incorrect answer cramps school buses and municipal vehicles while trying to maneuver through intersections with only a 35 ft inscribed radius. The inscribed radius is insufficient for WB-50 semi-trailer moving vans to maneuver through the roundabout without completely crossing the mountable center island. Snow storage during winter months will reduce the usable radius, requiring extra effort to remove the snow completely from the roundabout.

(D) This incorrect answer shows that a 45 ft inscribed radius is adequate for all expected types of vehicles. However, eliminating the center island allows stray semi-trailer rigs to pass straight through the intersection, removing the speed reduction and left-turn barrier benefits provided by a roundabout design.

SOLUTION 3

Crosswalk length is only one consideration when designing an intersection. The shortest length does not always result in the best location. MUTCD Sec. 3B.16.10 recommends placing the stop bar 4 ft ahead of the upstream side of the crosswalk, placing the crosswalk at the intersection. (See MUTCD Fig. 3B-18.)

The answer is (D).

Why Other Options Are Wrong

(A) This incorrect answer predicts that pedestrian interference will either stay the same or increase.

(B) This incorrect answer does not account for MUTCD's suggestion to locate a pedestrian yield sign 20 ft to 50 ft in advance of an unsignalized midblock crosswalk and locate a midblock crosswalk at least 100 ft from an intersection. Therefore, this is not a midblock crosswalk.

(C) This incorrect answer does not account for MUTCD's suggestion to locate a pedestrian yield sign 20 ft to 50 ft in advance of an unsignalized midblock crosswalk and locate a midblock crosswalk at least 100 ft from an intersection. Therefore, this is not a midblock crosswalk.

Intersection Geometry

5 Roadside and Cross-Section Design

PROBLEM 1

A paved parking area is being designed with $90°$ spaces that are 8.5 ft wide and 18 ft deep. The aisle is to be two-way and 24 ft wide. Drainage and maneuvering areas will make up 5% of the paved area.

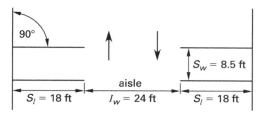

Approximately how many spaces can be provided per acre of pavement?

(A)   81 spaces/ac

(B)   120 spaces/ac

(C)   160 spaces/ac

(D)   170 spaces/ac

Hint: The area required for one space includes an adjoining section of aisle shared by another space.

PROBLEM 2

A parking lot is being designed for a capacity of 550 spaces. Spaces are to be 9 ft wide by 18 ft deep, placed at a $90°$ angle from a 22 ft aisle. Maneuvering areas and access driveways will occupy 3% of the parking area.

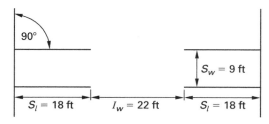

Federal regulations require that 2% of the total spaces be reserved for vehicles of disabled persons and that two of the handicapped spaces accommodate van parking.

The handicapped spaces must be 8 ft wide with an adjoining 5 ft access aisle.

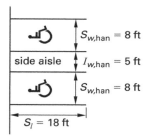

The handicapped van spaces must have an 8 ft access aisle. One access aisle may serve two abutting spaces.

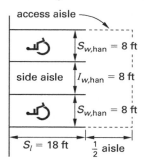

What is the approximate area required for the parking lot?

(A) $144,000 \text{ ft}^2$

(B) $149,000 \text{ ft}^2$

(C) $152,000 \text{ ft}^2$

(D) $208,000 \text{ ft}^2$

Hint: Assign one-half of the adjacent aisle width to the area required for each space.

PROBLEM 3

A 1.5 ac parking lot is to be laid out using the following configuration.

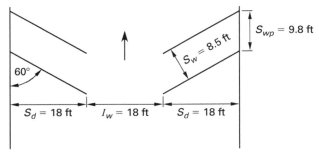

Driveways and maneuvering space require 10% of the total lot area. Most nearly, how many spaces will the lot accommodate?

(A) 110 spaces

(B) 220 spaces

(C) 240 spaces

(D) 250 spaces

Hint: A row module has the area of two spaces and the shared portion of the aisle between the spaces.

PROBLEM 4

An 80,000 ft^2 building is to be constructed with 45,000 ft^2 of daytime office space and 12,000 ft^2 of evening retail space. Mechanical space will make up 10% of the total area, and the remaining space will be for warehouse and storage use. The parking requirements described by a certain zoning code are as follows.

building space use	parking spaces required
office	1 space/300 ft^2
retail	1 space/150 ft^2
warehouse/storage	1 space/500 ft^2

Most nearly, how many parking spaces are required?

(A) 150 spaces

(B) 230 spaces

(C) 260 spaces

(D) 280 spaces

Hint: Each building space can be assigned only one type of use.

PROBLEM 5

A crash cushion is being designed for a 70 mph impact. The design criterion calls for a maximum deceleration rate of 7 g with 75% efficiency. Most nearly, what is the minimum compression length of the crash cushion?

(A) 15 ft

(B) 17 ft

(C) 23 ft

(D) 31 ft

Hint: The efficiency is comparable to a design factor of safety.

PROBLEM 6

An energy absorption barrier is being placed against a backwall. The barrier is to absorb the impact of a 4500 lbm vehicle approaching at a speed of 70 mph. Use a factor of safety of 1.5 and a deceleration rate no greater than 7.5 g. The vehicle is to stop in 23.4 ft.

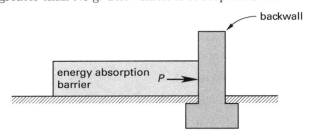

What is most nearly the force on the backwall?

(A) 32 kips

(B) 47 kips

(C) 51 kips

(D) 1500 kips

Hint: The impact force is the deceleration force.

SOLUTION 1

The number of spaces is determined by dividing the total lot area by the area required for one space, which includes a portion of the adjoining aisle needed for that space.

$$\text{no. of spaces} = \left(\frac{\text{total lot area}}{\text{area needed per space}} \right) \times (\text{proportion of area usable})$$

Determine the area required for two spaces and the adjoining center aisle.

$$\begin{aligned} A_2 &= S_w(2S_l + I_w) \\ &= (8.5 \text{ ft})\big((2)(18 \text{ ft}) + 24 \text{ ft}\big) \\ &= 510 \text{ ft}^2 \end{aligned}$$

Determine the area required for one space.

$$\begin{aligned} A_1 &= \frac{A_2}{2} = \frac{510 \text{ ft}^2}{2} \\ &= 255 \text{ ft}^2 \end{aligned}$$

Determine the net available area for parking after deducting for maneuvering and driveway space.

$$\begin{aligned} A_{\text{net}} &= A(\text{fraction of area usable}) \\ &= \left(43{,}560 \ \frac{\text{ft}^2}{\text{ac}} \right)(0.95) \\ &= 41{,}382 \text{ ft}^2/\text{ac} \end{aligned}$$

Determine the number of parking spaces available per acre.

$$\begin{aligned} \text{no. of spaces} &= \frac{41{,}382 \ \dfrac{\text{ft}^2}{\text{ac}}}{255 \ \dfrac{\text{ft}^2}{\text{space}}} \\ &= 162 \text{ spaces/ac} \quad (160 \text{ spaces/ac}) \end{aligned}$$

The answer is (C).

Why Other Options Are Wrong

(A) This answer incorrectly assumes that the initial area determination is for one space, not two.

(B) This answer incorrectly attaches to one space the full width of the adjacent aisle.

(D) This incorrect answer fails to reduce the available parking area by the maneuvering and aisle area.

SOLUTION 2

The total area required, A_{total}, is the sum of the areas required for each space required by type, A_i, multiplied by the number of spaces required by type, N_i. The sum is adjusted by a factor of utilization, f_u, which accounts for space needed for maneuvering and driveway areas.

$$A_{\text{total}} = \frac{\sum N_i A_i}{f_u}$$

Determine the area required for one normal space, A_n.

$$\begin{aligned} A_n &= S_w\left(S_l + \frac{I_w}{2} \right) = (9 \text{ ft})\left(18 \text{ ft} + \frac{22 \text{ ft}}{2} \right) \\ &= 261 \text{ ft}^2/\text{space} \end{aligned}$$

Determine the total handicapped spaces required, H.

$$\begin{aligned} H &= (\text{fraction of handicapped spaces required}) \\ &\quad \times (\text{no. of spaces provided}) \\ &= (0.02)(550 \text{ spaces}) \\ &= 11 \text{ spaces} \end{aligned}$$

Since two adjoining handicapped auto spaces may share the same side aisle, determine the area for two adjacent handicapped auto spaces, $A_{2,\text{han}}$.

$$\begin{aligned} A_{2,\text{han}} &= (S_{w,\text{han}} + I_{w,\text{han}} + S_{w,\text{han}})\left(S_l + \frac{I_w}{2} \right) \\ &= (8 \text{ ft} + 5 \text{ ft} + 8 \text{ ft})\left(18 \text{ ft} + \frac{22 \text{ ft}}{2} \right) \\ &= 609 \text{ ft}^2/2 \text{ spaces} \end{aligned}$$

Determine the area for one handicapped space with an access aisle, A_{han}.

$$\begin{aligned} A_{\text{han}} &= (S_{w,\text{han}} + I_{w,\text{han}})\left(S_l + \frac{I_w}{2} \right) \\ &= (8 \text{ ft} + 5 \text{ ft})\left(18 \text{ ft} + \frac{22 \text{ ft}}{2} \right) \\ &= 377 \text{ ft}^2/\text{space} \end{aligned}$$

Two handicapped van spaces, V_2, are required. Therefore, they can share the same access aisle.

Determine the area for the two handicapped van spaces, $A_{2,\text{van}}$.

$$
\begin{aligned}
A_{2,\text{van}} &= (S_{w,\text{van}} + I_{w,\text{van}} + S_{w,\text{van}})\left(S_l + \frac{I_w}{2}\right) \\
&= (8\ \text{ft} + 8\ \text{ft} + 8\ \text{ft})\left(18\ \text{ft} + \frac{22\ \text{ft}}{2}\right) \\
&= 696\ \text{ft}^2/2\ \text{spaces}
\end{aligned}
$$

Compute the total lot area required.

$$
A_{\text{total}} = \frac{NA_n + H_2 A_{2,\text{han}} + HA_{\text{han}} + V_2 A_{2,\text{van}}}{f_u}
$$

$$
\begin{aligned}
&= \frac{\begin{array}{l} (550\ \text{spaces} - 11\ \text{spaces})\left(261\ \dfrac{\text{ft}^2}{\text{space}}\right) \\[6pt] + \left(\begin{array}{l}11\ \text{handicapped spaces} - 2\ \text{van spaces} \\ -\ 1\ \text{handicapped space}\end{array}\right) \\[10pt] \times\left(\dfrac{609\ \text{ft}^2}{2\ \text{spaces}}\right) + (1\ \text{handicapped space}) \\[10pt] \times\left(\dfrac{377\ \text{ft}^2}{\text{space}}\right) + (2\ \text{van spaces})\left(\dfrac{696\ \text{ft}^2}{2\ \text{spaces}}\right) \end{array}}{0.97} \\[6pt]
&= 148{,}647\ \text{ft}^2 \quad (149{,}000\ \text{ft}^2)
\end{aligned}
$$

The answer is (B).

Why Other Options Are Wrong

(A) This incorrect answer is the total parking space and aisle area required without including the maneuvering and driveway area.

(C) This incorrect answer results from adding the number of handicapped spaces to the total required instead of including the handicapped spaces in the total.

(D) This incorrect answer results from including the entire aisle width adjacent to a space as the space requirement instead of using one-half of the aisle width.

SOLUTION 3

The number of spaces is found by dividing the total lot area by the area needed for each space plus the portion of the aisle needed for that space, then adjusting for driveways and maneuvering areas.

Determine the area required for a module of two spaces and the adjoining shared aisle.

$$
\begin{aligned}
A_{2,\text{sp}} &= S_{wp}(2S_d + I_w) \\
&= (9.8\ \text{ft})((2)(18\ \text{ft}) + 18\ \text{ft}) \\
&= 529.2\ \text{ft}^2
\end{aligned}
$$

Determine the area required for one space.

$$
\begin{aligned}
A_{\text{sp}} &= \frac{529.2\ \text{ft}^2}{2\ \text{spaces}} \\
&= 264.6\ \text{ft}^2/\text{space}
\end{aligned}
$$

Determine the total number of spaces possible.

$$
\begin{aligned}
\text{no. of spaces possible} &= \frac{\left(43{,}560\ \dfrac{\text{ft}^2}{\text{ac}}\right)(1.5\ \text{ac})}{264.6\ \dfrac{\text{ft}^2}{\text{space}}} \\
&= 247\ \text{spaces}
\end{aligned}
$$

Adjust for driveways and maneuvering space.

$$
\begin{aligned}
\begin{array}{l}\text{adjusted no. of} \\ \text{spaces possible}\end{array} &= (\text{no. of spaces possible}) \\
&\quad \times (\text{proportion of area usable}) \\
&= (247\ \text{spaces})(0.90) \\
&= 222\ \text{spaces} \quad (220\ \text{spaces})
\end{aligned}
$$

The answer is (B).

Why Other Options Are Wrong

(A) This incorrect answer results from using the aisle module unit area for one space instead of for two spaces.

(C) This incorrect answer results from failing to adjust for maneuvering and driveway area.

(D) This incorrect answer results from using the right-angle space width, S_w, which is the wrong dimension, to determine the module area required per space.

SOLUTION 4

The total number of parking spaces required, N_{total}, is the sum of spaces required for each assigned building use.

$$
N_{\text{total}} = N_o + N_r + N_{w/s}
$$

Determine the space assigned to mechanical uses, for which no parking spaces are required.

$$\begin{array}{l}\text{mechanical} \\ \text{space assigned}\end{array} = (80{,}000 \text{ ft}^2)(0.10)$$

$$= 8000 \text{ ft}^2$$

Determine the warehouse and storage space.

$$\begin{aligned}\text{space assignment} &= 80{,}000 \text{ ft}^2 - 45{,}000 \text{ ft}^2 \\ &\quad - 12{,}000 \text{ ft}^2 - 8000 \text{ ft}^2 \\ &= 15{,}000 \text{ ft}^2\end{aligned}$$

Determine the required number of parking spaces for offices.

$$\begin{aligned}N_o &= \frac{45{,}000 \text{ ft}^2}{300 \dfrac{\text{ft}^2}{\text{space}}} \\ &= 150 \text{ spaces}\end{aligned}$$

Determine the required number of parking spaces for retail use.

$$\begin{aligned}N_r &= \frac{12{,}000 \text{ ft}^2}{150 \dfrac{\text{ft}^2}{\text{space}}} \\ &= 80 \text{ spaces}\end{aligned}$$

Determine the required number of parking spaces for warehouse and storage use.

$$\begin{aligned}N_{w/s} &= \frac{15{,}000 \text{ ft}^2}{500 \dfrac{\text{ft}^2}{\text{space}}} \\ &= 30 \text{ spaces}\end{aligned}$$

Determine the total number of parking spaces required.

$$\begin{aligned}N_{\text{total}} &= 150 \text{ spaces} + 80 \text{ spaces} + 30 \text{ spaces} \\ &= 260 \text{ spaces}\end{aligned}$$

The answer is (C).

Why Other Options Are Wrong

(A) This is the number of spaces required for office spaces only. No specific information was given in the problem to allow consideration of office space and retail space having separate hours. Therefore, overlapping hours must be considered, requiring full accommodation of both.

(B) This incorrect answer does not include parking spaces needed for the warehouse and storage occupancy.

(D) This incorrect answer assigns parking spaces to the mechanical space as if it were warehouse and storage space. Mechanical space does not need parking space allocation.

SOLUTION 5

The deceleration distance is

$$s_d = \frac{\text{v}_{\text{final}}^2 - \text{v}_{\text{impact}}^2}{2a\eta}$$

$$= \frac{\left(0 \dfrac{\text{mi}}{\text{hr}}\right)^2 - \left(\left(70 \dfrac{\text{mi}}{\text{hr}}\right)\left(5280 \dfrac{\text{ft}}{\text{mi}}\right)\right)^2}{(2)(-7 \text{ g})\left(32.2 \dfrac{\text{ft}}{\text{sec}^2\text{-g}}\right)(0.75)\left(3600 \dfrac{\text{sec}}{\text{hr}}\right)^2}$$

$$= 31 \text{ ft}$$

The answer is (D).

Why Other Options Are Wrong

(A) This incorrect answer results from an improper conversion from miles per hour to feet per second.

(B) This incorrect answer results from inversely applying the efficiency factor.

(C) This incorrect answer results from ignoring the efficiency factor in the equation.

SOLUTION 6

The deceleration is derived from Newton's equation, with the factor of safety, FS, applied.

$$F = \frac{m}{g_c} a(\text{FS})$$

The stopping distance is

$$s = \frac{\text{v}_2^2 - \text{v}_1^2}{2a}$$

Rearrange to find a.

$$a = \frac{v_2^2 - v_1^2}{2s}$$

$$= \frac{\left(0 \; \frac{\text{mi}}{\text{hr}}\right)^2 - \left[\left(70 \; \frac{\text{mi}}{\text{hr}}\right)\left(5280 \; \frac{\text{ft}}{\text{mi}}\right)\right]^2}{(2)(23.4 \; \text{ft})\left(3600 \; \frac{\text{sec}}{\text{hr}}\right)^2}$$

$$= -225.2 \; \text{ft/sec}^2$$

The number of gravities is

$$\frac{-225.2 \; \dfrac{\text{ft}}{\text{sec}^2}}{32.2 \; \dfrac{\text{ft}}{\text{sec}^2\text{-g}}} = -6.99 \; \text{g} < 7.5 \; \text{g} \quad [\text{OK}]$$

Determine the force on the backwall.

$$F = \frac{m}{g_c} a(\text{FS})$$

$$= \left(\frac{4500 \; \text{lbm}}{32.2 \; \dfrac{\text{ft-lbm}}{\text{lbf-sec}^2}}\right)\left(-225.2 \; \frac{\text{ft}}{\text{sec}^2}\right)(1.5)$$

$$= 47{,}208 \; \text{lbf} \quad (47 \; \text{kips})$$

The answer is (B).

Why Other Options Are Wrong

(A) This incorrect answer does not apply the factor of safety to the deceleration force.

(C) This incorrect answer uses 7.5 g as the deceleration, whereas the problem called for the deceleration rate resulting from a 23.4 ft stopping distance.

(D) This is the result of not using the gravitational constant, g_c, to achieve consistent units between mass and force.

Roadside &
Cross-Section

6 Signal Design

PROBLEM 1

A three-phase signal with 4 sec lost time per phase has the following critical movement conditions.

	phase A	phase B	phase C
critical volume (vph)	70	400	550
adjusted saturation flow rate (vph)	350	1350	1960

Using the critical intersection volume-to-capacity ratio method, what is most nearly the recommended length of the cycle?

(A) 16 sec

(B) 27 sec

(C) 54 sec

(D) 81 sec

Hint: The optimal signal cycle, C_O, is one in which the critical intersection volume/capacity ratio, including lost time, is slightly less than 1. By inspection, all of the volume/service ratios shown are less than 1.

PROBLEM 2

The intersection of a minor street with a major two-lane, two-way street is being evaluated by city officials for the installation of a traffic signal. The major street meets the minimum *Manual on Uniform Traffic Control Devices* (MUTCD) warrant, but the minor street falls just short of the warrant. What additional information could help city officials justify the installation of a traffic signal at the intersection?

(A) A 15 min traffic jam occurs at the intersection when the employees of a nearby business leave work during the daily evening rush.

(B) An engineering study indicates that a traffic signal would improve the overall safety of the intersection.

(C) Five crashes at the intersection in the past year were caused by drivers not stopping at the stop sign on the minor street.

(D) One hundred people use the crosswalk at the intersection when traveling to work over a 2 hr morning period.

Hint: The motivation for installing a traffic signal needs to be supported by specific information.

PROBLEM 3

The geometric design and traffic for the intersection of 1st Ave. and Main St. are shown. Vehicle counts are for the peak one hour, and arrival is type 3 random. The posted speed limit is 35 mph, and there is a fixed time cycle length of 110 sec. The intersection is in a central business district (CBD), and there are no buses. The area served is urban, with a population of more than 250,000. The demand flow rate is 1900 pcphpl. The effective green time is 48 sec.

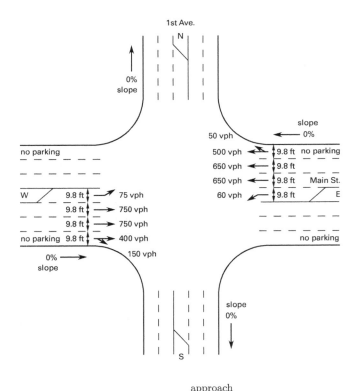

	approach					
	Main St. eastbound			Main St. westbound		
movement	left	through	through + right	left	through	through + right
max. green (sec)	8	50	50	8	50	50
min. green (sec)	8	50	50	8	50	50
yellow change (sec)	4	4	4	4	4	4
red clearance (sec)		2	2		2	2

Signal Design

If there is no initial queue, what is most nearly the capacity of the through-only movement on the east-bound approach?

(A)   720 pcph

(B)   1430 pcph

(C)   1500 pcph

(D)   1900 pcph

Hint: Through-only movements do not have turning traffic.

SOLUTION 1

The critical flow ratio, y_i, for each phase is the volume/ service flow ratio for the critical approach for each phase. V_i is the approach volume for the ith phase, and s_i is the adjusted saturation flow rate for the ith phase.

$$y_i = \frac{V_i}{s_i}$$

Solve for phase A.

$$y_A = \frac{V_A}{s_A} = \frac{70 \ \dfrac{\text{veh}}{\text{hr}}}{350 \ \dfrac{\text{veh}}{\text{hr}}}$$
$$= 0.200$$

Solve for phase B.

$$y_B = \frac{V_B}{s_B} = \frac{400 \ \dfrac{\text{veh}}{\text{hr}}}{1350 \ \dfrac{\text{veh}}{\text{hr}}}$$
$$= 0.296$$

Solve for phase C.

$$y_C = \frac{V_C}{s_C} = \frac{550 \ \dfrac{\text{veh}}{\text{hr}}}{1960 \ \dfrac{\text{veh}}{\text{hr}}}$$
$$= 0.281$$

The critical intersection volume/capacity ratio is the sum of the critical flow ratios for each phase.

$$\sum_{i \in ci} y_{c,i} = 0.200 + 0.296 + 0.281$$
$$= 0.777$$

The total cycle length can be estimated from HCM Eq. 19-30.

$$X_c = \left(\frac{C}{C - L}\right) \sum_{i \in ci} y_{c,i}$$

From *Highway Capacity Manual* (HCM) Eq. 19-1, the total lost time, L_t, for a three-phase signal with a 4 sec delay defaults to (3 phase)(4 sec/phase) = 12 sec. The critical intersection volume/capacity ratio should be no more than 1.0 to avoid oversaturated flow. After inserting known values, the equation becomes

$$\leq 1.0 = \left(\frac{C}{C - 12 \text{ sec}}\right) 0.777$$

By trial and error, the optimal cycle length is 54 sec.

The answer is (C).

Why Other Options Are Wrong

(A) In this incorrect answer, the critical intersection volume/capacity ratio included only the first two of the three approaches.

(B) In this incorrect answer, the lost time for each phase included only the start-up delay of 2 sec, and did not include the clearance delay.

(D) In this incorrect answer, the all-red interval default of 2 sec for each phase was added to the 4 sec delay for each phase. The 4 sec delay already includes the all-red interval.

SOLUTION 2

MUTCD is very clear that simply meeting the minimum warrants for a traffic signal is not sufficient to satisfy the requirements for installation. The *Traffic Engineering Handbook* further states that "...the requirements for a signal should be thoroughly analyzed with a decision to install based on a demonstrated traffic need."

Regardless of the minimum requirements that are met by traffic conditions, an engineering study is still needed to indicate that the installation will improve overall safety of the intersection.

The answer is (B).

Why Other Options Are Wrong

(A) A traffic jam in the evening caused by employee discharge from work can be reduced by alternate means, such as varying the quitting times of employees. Alternate means would have to be employed and observed before declaring that a signal is justified.

(C) Recurring crashes are only relevant if a traffic signal could somehow help prevent them. If drivers do not stop at a stop sign, there is no reason to believe they would stop at a traffic signal either. Thus, more information is needed about the nature of the crashes in order for this factor to be taken into consideration.

(D) Minimum pedestrian volume is to be 100 persons for each of any 4 hr period. Therefore, 100 persons for a 2 hr period is insufficient justification on its own.

SOLUTION 3

Solve for the following.

From *Highway Capacity Manual* (HCM) Eq. 19-9,

$$f_{HV} = \frac{100 - 0.3P_{HV} - 2.07P_g}{100}$$
$$= \frac{100 - 0 - 0}{100}$$
$$= 1.0$$

$f_w = 0.96$ [HCM Ex. 19-20]

$f_{bb} = 1.00$ [bus movements not given]

$f_a = 0.90$ [CBD default value, HCM p. 19-12]

$f_{LU} = 1.0$ [HCM p. 19-12]

$f_{LT} = 1.0$ [no turning movements in through-only lanes]

$f_{RT} = 1.0$ [no turning movements in through-only lanes]

$f_{Lpb} = 1.0$ [ped-bike counts not given]

$f_{Rpb} = 1.0$ [ped-bike counts not given]

Determine the saturation flow rate from HCM Eq. 19-8.

$$s_o = 1900 \text{ pcphpl}$$
$$s = s_o f_w f_{HV} f_p f_{bb} f_a f_{LU} f_{LT} f_{RT} f_{Lpb} f_{Rpb} f_{wz} f_{ms} f_{sp}$$
$$= \left(1900 \frac{\text{pc}}{\text{hr-ln}}\right)(0.96)(0.97)(1.0)(1.00)(0.90)(1.0)$$
$$\times (1.00)(1.0)(1.0)(1.0)$$
$$= 1592 \text{ pcphpl}$$

Determine the lane group volume-to-capacity ratio.

$$c = Ns\frac{g}{C} = (2)\left(1641 \frac{\text{pc}}{\text{hr-ln}}\right)\left(\frac{48 \text{ sec}}{110 \text{ sec}}\right)$$
$$= 1432 \text{ pcph} \quad (1430 \text{ pcph})$$

The answer is (B).

Why Other Options Are Wrong

(A) This incorrect answer is the capacity for one through-only lane.

(C) This incorrect answer is the total demand volume of the through lanes.

(D) This incorrect answer is the total through volume, including the volume from the shared right turn lane.

Signal Design

7 Traffic Control Design

PROBLEM 1

Two two-lane roads intersect at a four-way intersection as shown in illustration (a). To reduce the number of vehicle conflict points at the intersection, the intersection is to be replaced by the simple one-lane roundabout shown in illustration (b).

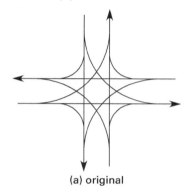

(a) original

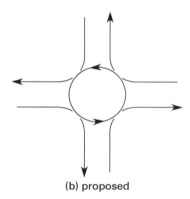

(b) proposed

Most nearly, how many crossing and merging conflict points will be eliminated by replacing the intersection with a roundabout?

(A) 8

(B) 12

(C) 20

(D) 28

Hint: A conflict occurs when two or more vehicle streams are able to occupy the same space at the same time.

PROBLEM 2

When a traffic signal is put in a flashing operation, which of the following statements is FALSE?

(A) A yellow indication should be used for the major street, and a red indication should be used for all other approaches.

(B) Each approach or separate controlled turn movement should be provided with a flashing display.

(C) Protected or protected/permissive left-turn movements should flash red when the through movement flashes yellow.

(D) The most restrictive minor street should have a red indication, and all other approaches may have a yellow indication.

Hint: A yellow indication does not require traffic to stop.

SOLUTION 1

There are 16 crossing conflicts, represented by black dots, and eight merging conflicts, represented by white dots, possible in the intersection configuration shown in illustration (a).

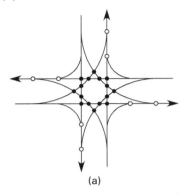

(a)

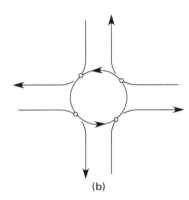

(b)

Roundabouts must be designed to allow the same movements to occur for each road. Roundabouts eliminate all left-turn conflicts. The roundabout design shown in illustration (b) has four merging conflicts and no left-turn or crossing conflicts.

The total number of conflicts eliminated by replacing the intersection with a roundabout is

$$N_t = 16 + 8 - 4 = 20$$

The answer is (C).

Why Other Options Are Wrong

(A) This incorrect answer does not consider sufficient left-turn conflicts and includes diverging conflicts in the total intersection count.

(B) This incorrect answer does not consider sufficient left-turn conflicts.

(D) This incorrect answer includes diverging conflicts in the total count.

SOLUTION 2

To avoid a possible moving conflict or collision, all signal faces should flash red, meaning all approaching traffic is to stop. Alternatively, a flashing yellow can be used for the major movement and for any nonconflicting turn movements that have adequate sight distance to proceed without coming to a full stop. It is unsafe to have more than one conflicting approach with yellow indication.

The answer is (D).

Why Other Options Are Wrong

(A) Allowing one approach (usually the major street) to proceed without stopping is permissible.

(B) Every movement that normally would be controlled with a R-Y-G signal must also be controlled with the red or yellow flashing signal.

(C) All flashing signal faces on an approach are to flash the same color, except for separate signal faces for protected or protected/permissive left-turn movements. These are to flash red when the through movement signal faces flash yellow (*Manual on Uniform Traffic Control* (MUTCD) Sec. 4D.30).

Geotechnical and Pavement

PROBLEM 1

During compaction of a parking area embankment under construction, a contractor reports the occurrence of pumping and rutting. In a standard field sand cone test, a 54 lbm soil sample with a volume of 0.39 ft³ was obtained. Using a field oven, the sample was dried to a new mass of 43 lbm. Given laboratory tests that indicate the specific gravity of the solids is 2.66, what is most nearly the original void ratio of the test sample?

(A) 0.2

(B) 0.3

(C) 0.5

(D) 2

Hint: The void ratio can be determined using a phase relationship because the total volume is known.

PROBLEM 2

Two 36 in diameter piles were designed for an allowable capacity on an individual basis, and were to be spaced 72 in apart. However, the two piles were mistakenly constructed 60 in apart. Given that the structural loading conditions, length of embedment, and soil conditions are the same for both piles, determine how the two piles will act.

(A) It cannot be determined, due to lack of pile dimensional information.

(B) It cannot be determined, due to lack of soil profile information.

(C) They will act in a group capacity.

(D) They will act in individual capacities.

Hint: Compare the circumference of the two piles as a group to the circumference of the two piles acting individually.

PROBLEM 3

A jurisdictional building code requires that maximum allowable pile loads be determined based on pile load testing data. The code specifies that the maximum allowable pile load is to be one-third the value of the test load that results in a total settlement value of not more than 0.002 in/kip. Given the pile load test data shown, use the code criteria to approximate the maximum allowable pile load.

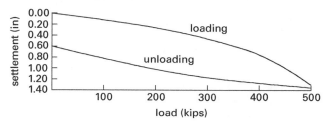

(A) 150 kips

(B) 230 kips

(C) 450 kips

(D) 500 kips

Hint: Use the maximum settlement criteria with the total load achieved during the pile load testing to determine the limiting settlement value and corresponding load.

PROBLEM 4

Which of the following statement(s) is/are true of the *Mechanistic-Empirical Pavement Design Guide* (MEPDG) design process?

I. The MEPDG process is a five-step, three-stage design and analysis process used to select a pavement strategy for a project.

II. The MEPDG process requires selecting a pavement design that can be standardized for an entire region based on a few select on-site tests, thereby reducing design costs.

III. The MEPDG process is only applicable to pavement rehabilitation projects and not for new project design.

IV. The MEPDG process can reduce the need to remove all of the existing pavement and subbase on a project that has failures only in selected locations or under selected local conditions.

V. The MEPDG process is completely based on new research in pavement design.

VI. The MEPDG process uses the International Roughness Index in relation to the design life as a part of the evaluation.

- (A) III only
- (B) IV and V only
- (C) I, II, and VI only
- (D) I, IV, and VI only

Hint: Pavement design has been documented in research and performance evaluations for more than 50 years, resulting in a multitude of reports, standards, and procedures.

SOLUTION 1

The void ratio is

$$e = \frac{(\text{SG}) \, V_t \rho_w}{m_s} - 1$$

$$= \frac{(2.66)(0.39 \text{ ft}^3)\left(62.4 \, \dfrac{\text{lbm}}{\text{ft}^3}\right)}{43 \text{ lbm}} - 1$$

$$= 0.5$$

The answer is (C).

Why Other Options Are Wrong

(A) This incorrect answer uses the total weight of the sample instead of the dry weight of the sample, producing a result that is less than the correct answer.

(B) This incorrect answer is actually the porosity determined by a simplified phase relationship equation (not including the percentage symbol). The result is less than the correct answer.

(D) This incorrect answer divides the total volume by the volume of the solids to calculate the void ratio. The value is greater than the correct answer and is the same as using the simplified equation and not subtracting 1.

SOLUTION 2

If the circumference of the two piles as a group is greater than the circumference of the two piles individually, the piles will retain 100% efficiency and no group reduction is required.

Calculate the group perimeter.

$$C_{\text{group}} = \pi D + 2(\text{spacing})$$
$$= \pi(36 \text{ in}) + (2)(60 \text{ in})$$
$$= 233.1 \text{ in}$$

Calculate the circumferential perimeter.

$$C_{\text{ind-pile} \times 2} = 2\pi D = 2\pi(36 \text{ in})$$
$$= 226.2 \text{ in}$$

The group perimeter is greater than the sum of the two individual circumferences. Therefore, the two piles will act individually.

The answer is (D).

Why Other Options Are Wrong

(A) This option is incorrect because the pile lengths are the same and because pile dimension information is not required, since the perimeter and circumference information is constant for each evaluation.

(B) This option is incorrect because soil profile information is not required, since the perimeter and circumference information is constant for each evaluation. Therefore, this determination can be made from a comparison of the required circumference length that would need to be mobilized to activate needed shear strength and achieve the proper friction capacity for the same loading conditions.

(C) This option mistakenly calculates the group perimeter by omitting the perimeter distance between the two piles and determining that the group perimeter is less than the sum of the two individual circumferences.

SOLUTION 3

The maximum settlement value is calculated using the code settlement criteria.

$$S_{max} = P_{test\,load,max}\left(0.002\ \frac{in}{kip}\right)$$
$$= (500\ kips)\left(0.002\ \frac{in}{kip}\right)$$
$$= 1\ in$$

Using the graphed data, the test load corresponding to the maximum settlement value of 1 in is

$$P_{test\,load,set} = 450\ kips$$

The maximum allowable pile load is

$$P_a = \tfrac{1}{3}P_{test\,load,set} = \left(\frac{1}{3}\right)(450\ kips)$$
$$= 150\ kips$$

The answer is (A).

Why Other Options Are Wrong

(B) This option mistakenly takes half the corresponding test load instead of one third.

(C) This option fails to reduce the corresponding test load by the required one third.

(D) This option mistakenly assumes the corresponding test load is the maximum test load.

SOLUTION 4

The MEPDG design process is an iterative five-step, three-stage process used to evaluate and predict multiple performance indicators. It specifies a method that integrates materials, structural design, construction, climate, traffic, and pavement management systems. The process can be customized to local conditions by modifying the process for local pavement distress and smoothness, as specified in supporting documents such as *Standard Practice for Conducting Local or Regional Calibration Parameters for the MEPDG* (NCHRP 2007.b). Pavement design is evaluated against design life, based on future pavement deterioration, including final threshold roughness. MEPDG can also be used to evaluate the condition of existing pavements scheduled for surface rehabilitation in order to avoid removing existing subbase and sub-pavement layers that are performing at an acceptable level.

The answer is (D).

Why Other Options Are Wrong

(A) This incorrect answer does not take into account that MEPDG is used for both pavement rehabilitation projects and new design.

(B) This incorrect answer does not take into account that MEPDG incorporates empirical procedures from *Guide for Design of Pavement Structures* (AASHTO, 1993) with multiple performance indicators, relying extensively on previously developed design procedures.

(C) This incorrect answer does not use extensive inputs of data from the entire region to determine performance-based design. The design will not be properly calibrated with only a few tests from selected on-site conditions.

Geotechnical and Pavement

9 Drainage

PROBLEM 1

A stormwater detention pond uses a submerged, sharp-edged orifice to control discharge to an open channel. A standpipe is used to prevent the pool elevation from exceeding 100 ft. The centerline of the orifice is at an elevation of 92.6 ft. Most nearly, what diameter of the orifice will limit the discharge to 20 ft^3/sec?

(A) 0.77 ft

(B) 0.85 ft

(C) 1.1 ft

(D) 1.4 ft

Hint: This is similar to discharge from an open tank.

PROBLEM 2

Manometers are installed on the upstream and downstream sides of a straight pipe 20 ft long with an inside diameter 0.75 in. The flow rate through the pipe is 8 gal/min. The upstream manometer reading is 37.1 in, and the downstream manometer reading is 9.2 in. Assuming turbulent flow, what is most nearly the specific roughness of the pipe?

(A) 0.000013 ft

(B) 0.00022 ft

(C) 0.014 ft

(D) 0.019 ft

Hint: Use the Darcy equation.

PROBLEM 3

A storm hydrograph is shown. The basin drainage area is 5480 mi^2.

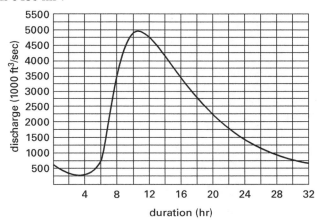

What is the approximate direct runoff from the storm?

(A) 0.016 in

(B) 2.6 in

(C) 13 in

(D) 16 in

Hint: Separate groundwater from direct runoff and integrate.

SOLUTION 1

Assume the orifice to have negligible loss from velocity or fluid contraction. The elevation head above the orifice is

$$h = 100 \text{ ft} - 92.6 \text{ ft}$$
$$= 7.4 \text{ ft}$$

For a sharp-edged orifice, the orifice coefficient is approximately 0.62.

$$A = \frac{Q}{C_d\sqrt{2gh}} = \frac{20 \; \dfrac{\text{ft}^3}{\text{sec}}}{(0.62)\sqrt{(2)\left(32.2 \; \dfrac{\text{ft}}{\text{sec}^2}\right)(7.4 \text{ ft})}}$$

$$= 1.48 \text{ ft}^2$$

Use the cross-sectional area equation and solve for the diameter.

$$A = \frac{\pi D^2}{4}$$

$$D = \sqrt{\frac{4A}{\pi}}$$

$$= \sqrt{\frac{(4)(1.48 \text{ ft}^2)}{\pi}}$$

$$= 1.4 \text{ ft}$$

The answer is (D).

Why Other Options Are Wrong

(A) This incorrect answer applies the square root only to the numerator in the final equation to determine diameter. Other assumptions, definitions, and equations are the same as those used in the correct solution.

(B) This incorrect answer multiplies instead of divides the flow rate by the orifice coefficient in the area equation. Other assumptions, definitions, and equations are the same as those used in the correct solution.

(C) This incorrect answer ignores the contraction and velocity losses corrected by the orifice coefficient. Other assumptions, definitions, and equations are the same as those used in the correct solution.

SOLUTION 2

Assume that the only head loss in the pipe is from friction.

$$h_f = z_2 - z_1 = \frac{37.1 \text{ in} - 9.2 \text{ in}}{12 \; \dfrac{\text{in}}{\text{ft}}}$$

$$= 2.33 \text{ ft}$$

The cross-sectional area of the pipe is

$$A = \pi \frac{D^2}{4} = \frac{\pi \dfrac{(0.75 \text{ in})^2}{4}}{\left(12 \; \dfrac{\text{in}}{\text{ft}}\right)^2} = 0.0031 \text{ ft}^2$$

The flow velocity is

$$v = \frac{Q}{A} = \frac{8 \; \dfrac{\text{gal}}{\text{min}}}{(0.0031 \text{ ft}^2)\left(60 \; \dfrac{\text{sec}}{\text{min}}\right)\left(7.48 \; \dfrac{\text{gal}}{\text{ft}^3}\right)}$$

$$= 5.8 \text{ ft/sec}$$

The friction factor is

$$f = \frac{2h_f D_g}{L v^2}$$

$$= \frac{(2)(2.33 \text{ ft})(0.75 \text{ in})\left(32.2 \; \dfrac{\text{ft}}{\text{sec}^2}\right)}{(20 \text{ ft})\left(5.8 \; \dfrac{\text{ft}}{\text{sec}}\right)^2\left(12 \; \dfrac{\text{in}}{\text{ft}}\right)}$$

$$= 0.014$$

Since flow is turbulent, use the von Karman-Nikuradse smooth pipe equation.

$$\frac{1}{\sqrt{f}} = 1.74 - 2\log\left(\frac{2\epsilon}{D}\right)$$

$$\frac{1}{\sqrt{0.014}} = 1.74 - 2\log\left(\frac{2\epsilon}{D}\right)$$

$$\frac{\epsilon}{D} = 0.00021$$

$$\epsilon = \left(\frac{\epsilon}{D}\right)D$$

$$= \frac{(0.00021)(0.75 \text{ in})}{12 \; \dfrac{\text{in}}{\text{ft}}}$$

$$= 0.000013 \text{ ft}$$

The answer is (A).

Why Other Options Are Wrong

(B) This incorrect answer calculates the relative roughness instead of the specific roughness. Other assumptions, definitions, and equations are the same as those used in the correct solution.

(C) This incorrect answer fails to convert units for friction loss from inches to feet. Other assumptions, definitions, and equations are the same as those used in the correct solution.

(D) This incorrect answer uses the wrong conversion factor for gal/min to ft^3/sec. Other assumptions, definitions, and equations are the same as those used in the correct solution.

SOLUTION 3

Separate groundwater from direct runoff as shown in the following illustration. Determine the direct runoff graphically.

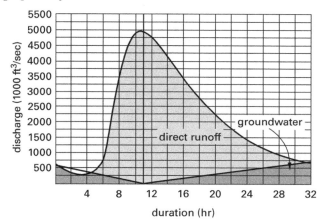

Each grid segment on the graph represents a discharge volume of

$$
\frac{\left(250{,}000 \ \dfrac{\text{ft}^3}{\text{sec}}\right)(2 \ \text{hr})\left(3600 \ \dfrac{\text{sec}}{\text{hr}}\right)}{43{,}560 \ \dfrac{\text{ft}^2}{\text{ac}}} = 41{,}322 \ \text{ac-ft}
$$

duration interval (hr)	segments
4–8	8.5
8–12	37
12–16	32.5
16–20	20
20–24	11.5
24–28	5.5
28–32	1.5
	116.5

The total discharge is

$$
Q = \frac{(116.5)(41{,}322 \ \text{ac-ft})\left(12 \ \dfrac{\text{in}}{\text{ft}}\right)}{(5480 \ \text{mi}^2)\left(640 \ \dfrac{\text{ac}}{\text{mi}^2}\right)}
$$

$$
= 16.47 \ \text{in} \quad (16 \ \text{in})
$$

The answer is (D).

Why Other Options Are Wrong

(A) This incorrect answer miscalculates the grid segment volume. The illustration and other assumptions, definitions, and equations are the same as those used in the correct solution.

(B) This incorrect answer uses the conversion from m^2 to ac instead of mi^2 to ac. The illustration and other assumptions, definitions, and equations are the same as those used in the correct solution.

(C) This incorrect answer improperly separates the groundwater flow from direct runoff. Other assumptions, definitions, and equations are the same as those used in the correct solution.

10 Alternatives Analysis

PROBLEM 1

A traffic signal system is to be installed at a shopping center entrance. The signal system costs $120,000 to install and is expected to have a salvage value of $15,000 at the end of its 15 yr life. The initial cost is to be depreciated using the straight-line method. If interest is 6%, what is most nearly the value of the system at the end of the tenth year?

(A) $40,000

(B) $51,000

(C) $61,000

(D) $70,000

Hint: The remaining value includes the undepreciated initial cost, plus the tenth-year salvage value.

PROBLEM 2

An escrow fund has been established in the amount of $2.5 million for demolition of a railroad crossing in 15 yr. The fund is expected to yield 4.5% interest compounded annually. The value of the fund at the end of the term will be most nearly

(A) $2,300,000

(B) $4,200,000

(C) $4,700,000

(D) $4,800,000

Hint: The nominal interest is the effective annual interest.

PROBLEM 3

A transit terminal in a small town is to be built using $500,000 of a bond issue. The bonds will yield 6% annually and will be retired at the end of 10 yr. A sinking fund is to be established to retire the bonds and pay the interest, into which 10 annual end-of-year payments will be made. The sinking fund will return a rate of 4.5% annually. The cost of selling the bonds is 2% of the issue. What is most nearly the annual amount necessary for payment into the sinking fund?

(A) $31,000

(B) $42,000

(C) $69,000

(D) $72,000

Hint: The cost of the bond issue is added to the principal amount.

SOLUTION 1

Determine the undepreciated amount at the end of the tenth year, starting with the total depreciation.

$$\text{total depreciation} = \text{installation cost} - \text{salvage value}$$
$$= \$120{,}000 - \$15{,}000$$
$$= \$105{,}000$$

$$\text{depreciation per year} = \frac{\text{total depreciation}}{\text{lifetime}}$$
$$= \frac{\$105{,}000}{15 \text{ yr}}$$
$$= \$7000/\text{yr}$$

$$\begin{aligned}\text{undepreciated value} &= \text{installation cost} - (10 \text{ yr})\\ \text{after 10 yr} &\quad \times (\text{depreciation per year})\end{aligned}$$
$$= \$120{,}000 - (10 \text{ yr})\left(\frac{\$7000}{\text{yr}}\right)$$
$$= \$50{,}000$$

The tenth-year salvage value will be the present value of the future $15,000, at 6% per year. The factor $(P/F, i\%, n)$ is determined from tables.

$$P = F(P/F, i\%, n) = F(P/F, 6\%, 5)$$
$$= (\$15{,}000)(0.7473)$$
$$= \$11{,}210$$

Determine the remaining value.

$$\text{value after 10 yr} = P + \text{undepreciated value after 10 yr}$$
$$= \$11{,}210 + \$50{,}000$$
$$= \$61{,}210 \quad (\$61{,}000)$$

The answer is (C).

Alternate Solution

$$P = F\left(\frac{1}{(1+i)^n}\right)$$
$$= (\$15{,}000)\left(\frac{1}{(1.06)^5}\right)$$
$$= \$11{,}210$$

Why Other Options Are Wrong

(A) This incorrect answer results from using the capital recovery method of determining the annual capital amount.

(B) This answer is incorrect because depreciation was taken to zero at the end of 15 yr, instead of to the salvage value.

(D) This answer is incorrect because the salvage value was taken as an increased value at year 10, assuming that the salvage value followed the interest rate back from the end-of-life amount to become a reversed increase in value.

SOLUTION 2

Use the following formula to find a future value of a present amount, compounded over 15 annual periods.

$$F = P(F/P, i\%, n) = P(1+i)^n$$
$$= (\$2{,}500{,}000)(1+0.045)^{15}$$
$$= \$4{,}838{,}200 \quad (\$4{,}800{,}000)$$

The answer is (D).

Why Other Options Are Wrong

(A) This is the amount of interest gained, not the total value of the fund.

(B) This incorrect answer results from adding simple interest to the initial amount instead of compounding interest.

(C) This incorrect answer results from reversing the interest and term values.

SOLUTION 3

Determine the value of the bond reissue, F.

$$F = (\$500{,}000)(1.02) = \$510{,}000$$

Determine the annual bond yield, which is the product of the annual bond interest and value of the bond reissue.

$$\text{annual bond yield} = (\text{annual bond interest})F$$
$$= (0.06)(\$510{,}000)$$
$$= \$30{,}600$$

Determine the sinking fund amount, A, using the following formula.

$$A = F\left(\frac{i}{(1+i)^n - 1}\right)$$
$$= (\$510{,}000)\left(\frac{0.045}{(1+0.045)^{10} - 1}\right)$$
$$= \$41{,}503$$

Determine the annual amount necessary for payment into the sinking fund, which is the sum of the annual bond interest plus the sinking fund amount, A.

$$\begin{aligned}
\text{annual cost} &= \text{annual bond yield} + A \\
&= \$30{,}600 + \$41{,}503 \\
&= \$72{,}103 \quad (\$72{,}000)
\end{aligned}$$

The answer is (D).

Why Other Options Are Wrong

(A) This incorrect answer is the annual cost of the bond interest only.

(B) This incorrect answer is the annual cost of the sinking fund necessary to pay off the bond face value only.

(C) This incorrect answer assumes that the 2% bond sales cost is not a financable amount but is deducted from the bond face value.

Alternatives Analysis